TOPOLOGY
An Introduction to the Point-Set
and Algebraic Areas

TOPOLOGY
An Introduction to the Point-Set and Algebraic Areas

Donald W. Kahn
Professor of Mathematics
University of Minnesota

DOVER PUBLICATIONS, INC.
New York

Copyright

Copyright © 1995 by Donald W. Kahn.
Copyright © 1975 by The Williams & Wilkins Company.
All rights reserved under Pan American and International Copyright Conventions.

Published in Canada by General Publishing Company, Ltd., 30 Lesmill Road, Don Mills, Toronto, Ontario.
Published in the United Kingdom by Constable and Company, Ltd., 3 The Lanchesters, 162–164 Fulham Palace Road, London W6 9ER.

Bibliographical Note

This Dover edition, first published in 1995, is an unabridged, corrected and slightly enlarged republication of the work first published by The Williams & Wilkins Company, Baltimore, 1975. For the Dover edition the author has provided a new section, "Solutions to Selected Problems."

Library of Congress Cataloging-in-Publication Data

Kahn, Donald W., 1935–
    Topology : an introduction to the point-set and algebraic areas / Donald W. Kahn.
    p. cm.
    "An unabridged, corrected and slightly enlarged republication of the work first published by the Williams & Wilkins Company, Baltimore, 1975. For the Dover edition the author has provided a new section, 'Solutions to selected problems'"—T.p. verso.
    Includes bibliographical references (p.   -   ) and index.
    ISBN 0-486-68609-4
    1. Topology. I. Title.
QA611.K32 1995
514'.2—dc20
                                              95-1157
                                                 CIP

Manufactured in the United States of America
Dover Publications, Inc., 31 East 2nd Street, Mineola, N.Y. 11501

Preface

What could be the justification for a new text in beginning topology? We have seen, in recent years, the appearance of many creditable topology texts; in point-set topology, there are fine books by Kelley, Gaal, Bourbaki, and others, while in algebraic topology, such books as those by Massey, Spanier, and Dold have brought large parts of the subject into a rather definitive form. No one can possibly instill much originality into these topics, already perfected over the years. Nevertheless, some very practical needs are unattended.

Here at the University of Minnesota, we have a basic curriculum in topology consisting of an introductory course in general and algebraic topology—for seniors and beginning graduate students—to be followed by a specialized graduate course in homology and homotopy theory. The subject matter of the beginning course, which is the subject matter of this text, was worked out through some (often delicate) negotiations, in committee. It was agreed that—at this level—we should offer two quarters of general topology, to be followed by one quarter of algebraic topology, specifically 2-manifolds, covering spaces and fundamental groups. With respect to this curriculum, the textbook situation is very poor. To cover the material, two or more books had to be used, sometimes with each being employed for a small part of the course. Apart from the added burden on the students, who buy the books, the use of small, sometimes isolated parts of a text is often confusing. In general topology, in particular, there are often many technical notions, which can in no way help a beginner to grasp the fundamentals of the subject. The use of small portions of a comprehensive text in general topology is therefore a difficult matter.

The purpose of this book is then entirely practical. We wish to make available, in a single book, at as moderate a price as possible, the curriculum in beginning topology. While I can't say that our particular choice of subject matter is identical with that at most other schools, I hope the text will prove to be of general use.

I have adhered to several principles:

1. The text is designed to be read—in its entirety—by any student at the proper level (who is conversant in the basics of real analysis or advanced calculus).

2. The problems are all intended to assure a grasp of the materials; there are absolutely no tricks, nor problems which range far afield. It is my serious intention that an industrious student should try them *all*. A few problems contain material used at a later point; this is noted in the text, and is generally kept at a minimum.

3. A student often wishes that the author of his/her texts supplies absolutely all details. I have found this unsatisfactory, for the text becomes cumbersome and—more importantly—the student never learns to think properly. The other extreme, where the author merely gives a sketch, is equally useless, as the reader becomes too frustrated to make any progress. Obviously, a moderate stand is needed, but I have tried to point out—in the beginning chapters—where some details should be supplied by the reader. (Frequently, I write "check this.")

4. I consider the most important thing here to be frankness. I have made every effort to say clearly and honestly, where the student can find topics which I have had to omit, or a source of problems more challenging than those here. Like every author, I have made choices. For example, in the area of separation axioms in topology, I consider the Hausdorff condition and normality to be basic. Such refinements as complete regularity or full normality have no natural place in such a beginning course, and any industrious student can easily track them down from the Bibliography.

Similarly, on occasion I give a complete treatment of a special case, followed by a complete treatment of a more general case at a later moment. Of course, efficiency is generally desirable, but only a fool would be so doctrinaire as to allow efficiency to destroy a clear and reasoned presentation of a subject.

These matters quickly evolve into questions of taste, and only the reader can judge whether I have succeeded.

Finally, let me note that which is obvious to all authors. One formulates ones ideas in an elaborate manner, often under the influences of many people. It's never easy to point out all forces which led to the ideas in this book. The origins of theorems can frequently be traced in the references, but the selection of ideas and the method of presentation—often coming from my teachers and friends—are hard to credit precisely. I am grateful to them all, while, of course, accepting full responsibility for everything here.

Contents

Preface... v

Chapter 1. Logic, Set Theory, and the Axiom of Choice........ 1
This review chapter surveys elementary logic and set theory, including such concepts as countability, power set, etc. The axiom of choice is presented, along with its most useful variants. The chapter closes with an outline of category theory.

Chapter 2. Metric Spaces.................................... 15
After a review of the real numbers, Euclidean spaces are studied in detail. This is followed by a treatment of all the basic properties of metric spaces and continuous functions from a metric space to another.

Chapter 3. General Topological Spaces. Bases. Continuous Functions. Product Spaces........................... 35
General topological spaces are treated here. Concepts such as interior, exterior, boundary, closure, etc., are examined in detail. Bases and subbases are set up, and the relative and product topologies are defined. After a treatment of continuous functions, the chapter closes with the notion of a homeomorphism.

Chapter 4. The Special Notions of Compactness and Connectedness... 55
The Hausdorff separation axiom is introduced here; compact spaces are examined from various viewpoints, and the compact subspaces of a Euclidean space are found. More, generally, compact metric spaces are characterized by the Bolzano-Weierstrass property. The Tychonoff theorem is then proved. The chapter ends with a study of the basic properties of connected and arcwise connected spaces.

Chapter 5. Sequences, Countability, Separability, and Metrization.. 77
The first and second axioms of countability are introduced, as well as the notion of separability. The role of sequences in the topology of first countable spaces is established. Normal spaces are treated, followed by Urysohn's lemma, and finally the Urysohn metrization theorem.

Chapter 6. Quotients, Local Compactness, Tietze Extension, Complete Metrics, Baire Category.................... 92

This chapter surveys a variety of topics which are not in the main line of the earlier presentation, but most of which are applied later in the book. Included at this point are such topics as one-point compactifications and the uniform convergence of functions.

Chapter 7. Generalities about Manifolds; The Classification of Surfaces... 105

Manifolds are defined, and the basic properties, including an elementary imbedding theorem, are proved. Projective spaces and spheres are the beginning examples. Topological groups, Lie groups, and differentiable manifolds are briefly surveyed. The remainder of the chapter is devoted to a classification of compact surfaces.

Chapter 8. The Fundamental Group....................... 138

This is the first time where an algebraic invariant is introduced to study topological problems. The construction and basic properties occupy most of the chapter. At the end, the fundamental group is studied for Cartesian products and topological groups.

Chapter 9. Covering Spaces............................... 155

Covering spaces are defined, and the questions of lifting paths and homotopies are treated. This yields the classification of covering spaces in terms of subgroups of the fundamental group. The existence of covering spaces for semi-locally simply-connected spaces is next. The chapter ends with some computational examples.

Chapter 10. Calculation of Some Fundamental Groups. Applications.. 180

The fundamental group is studied for a connected simplicial complex. After some preliminaries, the fundamental group is proven to be isomorphic to the edge-path group. Various applications and results on the general structure of the fundamental group follow.

Epilogue... 203

Bibliography.. 207

Index of the Most Common Symbols...................... 208

Solutions to Selected Problems............................ 209

Index.. 215

CHAPTER 1

Logic, Set Theory, and the Axiom of Choice

Our purpose in this introductory chapter is to review some of the foundation concepts, on which the remainder of the text rests. We assume that someone who wishes to learn topology has some familiarity with all of these ideas, but we also include here some fairly extensive references so that a prospective student may fill in, on his own, any gaps in his knowledge. A basic tool in mathematics, at virtually any level, is logic. While logic has been developed into an elaborate subject of its own, often under the name of symbolic logic or mathematical logic, the logic which we need here deals mostly with the basic rules for manipulation of mathematical statements. This is a sensible place to begin our review; we indicate some of the more fancy treatments in the References at the end of the chapter.

One traditionally refers to mathematical statements by capital letters, such as P, Q, etc. Thus P can refer to things like

"49 is a positive number,"
"there are infinitely many points in the plane,"
"the area of a circle is πr^2,"
etc.

As a matter of convention, when one writes a mathematical statement, by itself, we mean to say that the statement is true. For example, if we wrote "Professor Schmitt has proved P in his last lecture. Hence, we conclude Q," we really mean that Schmitt has shown that P is true and thus we conclude (somehow, presumably by other remarks) that statement Q is true. Or another way this may arise is that we give the name P to some statement; after several lines, we show that P is true, but we write, for short, something like "Hence P."

On the other hand, when mathematical statements or propositions occur in conjunction with other statements, they may or may not be true. The standard sort of expression, which occurs all the time, is

"If P, then Q."

This really means, if P is true, then Q is true. It is an accepted convention that when P is false, the total assertion "If P, then Q" is regarded as true because if P is not true, we need not worry about the truth of the claim "if P is true, then Q is true." For example, the ridiculous assertion

"If $2 + 2 = 5$, then $3 + 3 = 7$"

is regarded as correct or true, because since $2 + 2$ is not 5, it doesn't matter what $3 + 3$ is.

A standard and convenient notation is $\sim P$, meaning not P or the negation of P. If P means x is greater than 3, then $\sim P$ would mean x is less than or equal to 3. Standing by itself, $\sim P$ would mean that P is false. Clearly, $\sim\sim P$ means P.

Various notions have become accepted. For example, "P and Q" means that both P and Q are true. "P or Q" means that at least one of the two statements is true. It is important to note that when we say "P or Q" we do allow the possibility that both are true. If we wish to exclude this case, we would write "P or Q" but not "P and Q."

One can now express some of these concepts in simpler and more concise ways. As a simple illustration, "P or Q" is precisely the same thing as $\sim(\sim P$ and $\sim Q)$, because "P or Q" encompasses all possibilities except that both P and Q are false. As a deeper application of this idea, we can rewrite "if P, then Q" as $\sim(P$ and $\sim Q)$, because it really means that if P is true, so is Q, which is the same thing as saying that it is not the case that if P is true, Q is false.

The statement "if P, then Q" occurs in various costumes all of which mean the same thing:
"when P is true, so is Q"
"P implies Q"
"under the hypothesis P, we have the conclusion Q"
"P is a sufficient condition for Q"
etc.
We will use the standard shorthand "$P \Rightarrow Q$" for "if P, then Q." Note that if "$P \Rightarrow Q$" is true, it need not follow that "$Q \Rightarrow P$" is true. If $P \Rightarrow Q$ and $Q \Rightarrow P$, then we say that P and Q are equivalent (the being both true or both false). This is usually written $P \equiv Q$ and is sometimes referred to as "a necessary and sufficient condition for Q is P," or "P if and only if Q."

In more complicated situations, P, Q, etc., depend on variables and are not absolute statements. For example, "$x > 5$" is a statement which is true when x is bigger than 5 but is false otherwise. In general, the dependency of statements on variables means that things may or may not be true, but it does not affect the rules of logic which deal with manipulating statements.

The basic rules, which may be easily checked, are the following:
1) $P \Rightarrow P$
2) If $P \Rightarrow Q$ and $Q \Rightarrow R$, then $P \Rightarrow R$.
3) $P \Rightarrow Q$ is equivalent to $\sim Q \Rightarrow \sim P$.
4) $\sim(P \text{ and } \sim P)$.

If we wanted to prove 3) we would show that
 a) if $P \Rightarrow Q$, then $\sim Q \Rightarrow \sim P$
and
 b) if $\sim Q \Rightarrow \sim P$, then $P \Rightarrow Q$.

For illustration, we write out the proof of a).

We suppose $P \Rightarrow Q$. We wish to show that if $\sim Q$ is true, then $\sim P$ is true. So we assume also that $\sim Q$ is true or Q is false. The two statements $P \Rightarrow Q$ (which is really $\sim(P \text{ and } \sim Q)$) and $\sim Q$ require that P is false, because if P were true, as $\sim Q$ is true, then we would have "P and $\sim Q$." But we know that "P and $\sim Q$" is false. Thus, P is false, or equivalently, $\sim P$ is true, which is what we wanted.

All the other assertions above are proved by equally elementary reasoning and we assume the student can handle them without difficulty. One frequently reasons, as we just did, that something must be true because its negation or opposite is not possibly true. Written out in all its formality, this is sometimes referred to as "proof by contradiction." It simply means that to prove P, one assumes for a moment that $\sim P$ is true, and then reasons that this cannot possibly be the case, so that P had to be true.

The theory of sets underlies virtually every branch of mathematics. Our ideal is to give a brief and intuitive review of the subject. The elaborate treatment is nowadays a branch of mathematical logic. A set is a collection of elements and is a concept which is familiar to virtually every student. Examples are
 a) the set of real numbers
 b) the set of points in the plane
 c) the set of whole numbers
 etc.

The basic concept here is that of an element or a member of a set. A set is really made up of its members or elements, and, to know a set, it suffices to know precisely what are its elements. This has given rise to a special notation which is universal in mathematics; we write "$x \in S$" to mean that x is an element of the set S. For example if T is the set of even numbers then "$2 \in T$" is a correct or true statement. Similarly, one writes $x \notin S$ when the element x does not belong to the set S. Thus, in our example, we would have $3 \notin T$.

4 Topology

The basic concepts in set theory are modeled on the concepts of logic given earlier in this chapter. We make the following definitions:

1) $S \subseteq T$ is defined by $(x \in S) \Rightarrow (x \in T)$; that is to say, if x belongs to S, it also belongs to T. We use the parenthesis to keep the different parts of the statement clearly separated from one another. Intuitively, $S \subseteq T$ means that S is a part of T or that S is contained in T. If S is the positive numbers and T is all the numbers, then clearly $S \subseteq T$. Note that we always have $S \subseteq S$, with this definition. We say that S is a *subset* of T.

2) We use the notation $\{x \mid P \text{ is true}\}$ to mean those elements x which make the statement P true. For example, the positive numbers could be written $\{x \mid x \text{ is a number and } x > 0\}$.

3) $S \cup T$ is defined as $\{x \mid x \in S \text{ or } x \in T\}$. It is called the *union* of S and T and means all those elements which belong to S or T (or possibly both).

4) $S \cap T$ is defined to be $\{x \mid x \in S \text{ and } x \in T\}$. It is called the *intersection* of S and T, and means all those elements belonging to both S and T.

These concepts can be illustrated graphically if we take as our sets, some sets of points in the plane. For example, the figures a), b) and c).

a)

b)

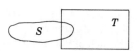

(The entire figure represents $S \cup T$)

c)

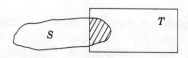

(The shaded area represents $S \cap T$)

represent subset, union, and intersection.

In addition, if $S \subseteq T$, one defines the difference $T - S = \{x \mid x \in T,$

$x \notin S\}$. It has proven useful to define a set with no elements, the *empty set*, usually written ϕ. Clearly, for any set S, $\phi \subseteq S$, because the condition for $\phi \subseteq S$ requires that every element of ϕ belongs to S, and as ϕ does not have any elements, this condition is in fact satisfied. Similarly, if S and T have no elements in common, we have

$$S \cap T = \phi.$$

Just as in calculus, one defines functions or maps $f: S \to T$ from one set to another. Such an f assigns to each element $x \in S$, a single element $f(x) \in T$. We have the usual definitions:

1) f is one-to-one (1-1), if whenever $f(x) = f(y)$, we have $x = y$. That is no two different elements are mapped to the same element.

2) f is onto, if every $y \in T$ is of the form $f(x)$, for some (possibly many) $x \in S$.

3) If f is 1-1 and onto, we say that f is a 1-1 correspondence (or sometimes we say that f is an isomorphism of sets). If f is a 1-1 correspondence from S to T, we may define $g: T \to S$ by $g(y)$ is that unique x so that $f(x) = y$. g is an inverse function to f and is trivially checked to be a 1-1 correspondence from T to S. We write $g = f^{-1}$ for this situation.

If there is a 1-1 correspondence, $f: S \to T$, S and T may be regarded as having the same number of elements (in classical set theory, one often says that S and T have the same cardinality). Some remarkable examples of 1-1 correspondence are now easy to clarify.

a) The even numbers (positive, negative, and zero) and all the whole numbers have the same cardinality. $f(n) = n/2$ gives the 1-1 correspondence (it makes sense because n is even). Note that we could just as easily define $f^{-1}(m) = 2m$, getting a map from all whole numbers to the even ones.

b) The even numbers and the odd numbers have the same cardinality. Define $f(n) = n + 1$.

c) The positive integers (whole numbers), have the same cardinality (or are in 1-1 correspondence with) the rational numbers (that is the fractions p/q with p and q whole numbers and q not zero). To prove this, we write the rational numbers as a table in the fourth quadrant.

$$0, 1, 2, 3, \cdots$$
$$-1, -2, -3, -4, \cdots$$
$$\tfrac{1}{2}, \tfrac{3}{2}, \tfrac{5}{2}, \cdots$$
$$-\tfrac{1}{2}, -\tfrac{3}{2}, -\tfrac{5}{2}, \cdots$$
$$\tfrac{1}{3}, \tfrac{2}{3}, \cdots$$
$$\vdots \quad \vdots \quad \vdots$$

We now construct a path which runs through this quadrant and touches each place exactly once.

6 Topology

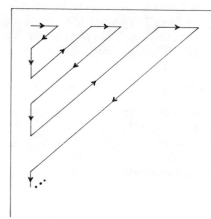

We define a function from the rational numbers to the positive integers by assigning to each rational number, the place it has, when one counts the numbers as they occur on this path. Because different numbers lie in different places, the map or function is 1-1. As every place is occupied by some number, the map is onto.

d) The positive whole numbers are in 1-1 correspondence with all the whole numbers. We put f: (positive whole numbers) → (all whole numbers)

$$f(1) = 0$$

$$f(2n - 1) = n - 1, \quad \text{whenever} \quad n > 1$$

$$f(2n) = -n, \quad \text{whenever} \quad n \geq 1.$$

One can check quickly that this procedure defines a function which is both 1-1 and onto.

e) Two finite sets are in 1-1 correspondence, whenever they have the same number of elements in the usual sense. If T is the whole numbers and S is the non-zero whole numbers, we can form a 1-1 correspondence $f: T \to S$ by

$$f(n) = n, \quad \text{if} \quad n < 0$$

$$f(n) = n + 1, \quad \text{if} \quad n \geq 0.$$

f) Not all infinite sets are in 1-1 correspondence. G. Cantor first proved that there is *no* 1-1 correspondence between the positive integers (whole numbers) and the real numbers (which we think of as decimals). For suppose there were such a correspondence; we write it as a table.

n	$f(n)$
1	$a_1 a_2 \cdots a_k\, a_{k+1}\, a_{k+2} \cdots$
2	$b_1 b_2 \cdots$
3	$\cdots$
$\vdots$	$\vdots\quad\vdots\quad\vdots$

Create a new decimal according to the following procedure. In the first place, choose a number different from a_1, in the second place a number not b_2, etc. This gives rise to a new decimal, not in the list, proving that f could not have been onto. In particular, f could not have been a 1-1 correspondence.

(Actually, this is a bit vague, as are decimals themselves. For example,

$$12.300\cdots = 12.2999\cdots.$$

We leave it to the reader to specify conventions to make this a precise proof, actually an easy task.)

A set which is in 1-1 correspondence with the integers (whole numbers) is called *countable* or *denumerable*. We have shown that the rational numbers are countable, while the real numbers are not. A remarkable construction, called the *power set*, always allows us to pass from a set S to a bigger set. We define the power set of S, written $P(S)$ as the set whose elements are the subsets of S (including the empty set). For example, $P(\phi)$ is a set with one element. If S has three elements, one easily checks that $P(S)$ has eight elements.

We show that there is never a 1-1 correspondence $f: S \to P(S)$. For suppose there were such an f. Define a set

$$U = \{x \mid x \in S,\ x \notin f(x)\}$$

This makes sense because $f(x)$ is really a subset of S. Note that $U \subseteq S$ and that $U \in P(S)$, i.e. U is an element of the set of all subsets of S.

If f were onto, there would be $y \in S$ so that $f(y) = U$. There are then two possibilities, either $y \in U$ or $y \notin U$. We will show both to be false, so our initial assumption (that there was such an f) had to be false.

If $y \in U$, $y \in S$, $y \notin f(y)$. But $f(y) = U$, so $y \in U$ and $y \notin U$, which is impossible.

If $y \notin U$, then $y \notin f(y)$. As $y \in S$, we have, by definition, $y \in U$, also a contradiction, completing the proof.

One concludes that there is no meaningful concept of a largest set, whereas ϕ is clearly a well-defined smallest set. More specific, we introduce a notion of smallness in sets by stipulating that S is smaller than or equal T if S is in 1-1 correspondence with some subset of T. In this context, a natural question is the question of whether the assumptions that S is smaller than

8   Topology

or equal to T and T is smaller than or equal to S imply that S and T are in 1-1 correspondence. This was answered affirmatively by Bernstein, Cantor, and Schroder, and we give now a brief proof.

Suppose S is in 1-1 correspondence with a subset of T and vice versa. There are then 1-1 (but not necessarily onto) maps

$$f \colon S \to T$$

$$g \colon T \to S.$$

Given $x \in S$, we may or may not have $y_1 \in T$ so that $g(y_1) = x$. If such a y_1 exists, we say that x pulls back once. Of course, there is only one y_1, since g is 1-1. If there is x_2 with $f(x_2) = y_1$, then $g(f(x_2)) = x$ and we say that x pulls back twice. In general, an element may pull back a finite or possibly an infinite number of times. With this in mind, we define,

$$S_{\text{even}} = \{x \mid x \in S, \ x \text{ pulls back an even number of times}\}$$

$$S_{\text{odd}} = \{x \mid x \in S, \ x \text{ pulls back an odd number of times}\}$$

$$S_{\infty} = \{x \mid x \in S, \ x \text{ pulls back an infinite number of times}\}.$$

In the exact same way, we can define T_{even}, T_{odd}, and T_{∞}. Now S_{even}, S_{odd}, and S_{∞} are all subsets of S, and they are disjoint, that is they have no element in common. Note that we regard 0 as an even number, so that the elements which do not pull back at all belong to S_{even}. Clearly, every element belongs to one of these three sets, that is

$$S = S_{\text{even}} \cup S_{\text{odd}} \cup S_{\infty}.$$

The same remarks obviously apply to T.

It is easy to check that f maps S_{even}, in a 1-1 way, onto T_{odd}. For f is 1-1, and if $y \in T_{\text{odd}}$, y pulls back an odd number of times, and hence at least once (because 1 is the smallest odd number). Similarly, g maps T_{even} onto S_{odd}, so that g^{-1} maps S_{odd} onto T_{even} in a 1-1 fashion. Finally f maps S_{∞} onto T_{∞} because if an element pulls back infinitely many times, it pulls back once to some other element which must also pull back infinitely many times.

We now define a 1-1 correspondence $h \colon S \to T$ by the following rule

$$h(x) = \begin{cases} f(x), & \text{if } x \in S_{\text{even}} \\ g^{-1}(x), & \text{if } x \in S_{\text{odd}} \\ f(x), & \text{if } x \in S_{\infty}. \end{cases}$$

This means that we define h by these different rules, depending on which case we are in. h is clearly 1-1, because the separate cases are disjoint sets, which are mapped to disjoint sets by 1-1 maps. h is onto because every element of T is in T_{odd}, T_{even}, or T_{∞}, and in each case, the map which we use to define h is onto. This situation is best portrayed by the following picture

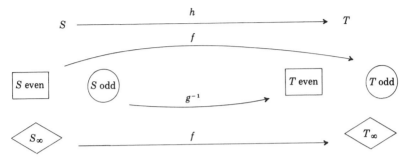

where the ingredients which make up h are displayed in detail under the map h.

There are other important constructions on sets, such as the Cartesian product of sets (named for R. Descartes, the French geometer and philosopher). This represents an abstract construction which generalizes the construction of the coordinate plane. Let S and T be sets. We define $S \times T$, the Cartesian product of S and T to be the set of all pairs (x, y), where $x \in S, y \in T$. That it is the set of all pairs of elements, the first from S and the second from T. We may define maps

$$P_1 : S \times T \to S \quad \text{by} \quad P_1((x, y)) = x$$

and

$$P_2 : S \times T \to T \quad \text{by} \quad P_2((x, y)) = y.$$

A map $h: X \to S \times T$, where X is any set, is completely determined by the two composite maps $P_1 \cdot h$ and $P_2 \cdot h$ (that is $P_1(h(x))$ and $P_2(h(x))$ for any $x \in X$).

It is also useful to define the Cartesian product of a family of sets (more than two). Suppose S_α represents a family of sets, one for each α (not necessarily finite in number). The union of all the S_α, written

$$\bigcup_\alpha S_\alpha,$$

is defined to be the collection of all elements which belong to at least one of the S_α. The Cartesian product of the S_α, written

$$\underset{\alpha}{\times} S_\alpha,$$

is defined to be the set of all functions

$$f: A \to \bigcup_\alpha S_\alpha,$$

where A is the set of all the α's and f is required to have the property that $f(\beta) \in S_\beta$ for each $\beta \in A$. One should check that this definition is the same as the above definition for the case where A has two elements. This concept will play an important role in point-set topology (later chapters).

A final review topic in set theory is the axiom of choice. This is an axiom in set theory, which has been shown (P. Cohen) to be independent of the other axioms, and which has been used, virtually as a matter of tradition in nearly all modern mathematics. It has many equivalent forms, also frequently used, and our goal here is to explore some of these variants. Stated briefly, we have

Axiom of Choice

Let S_α be a collection of non-empty sets, one for each $\alpha \in A$. Then, there is a function
$$f: A \to \bigcup_\alpha S_\alpha$$
with $f(\beta) \in S_\beta$ for every $\beta \in A$.

It is obvious, from the definition, that the axiom of choice is equivalent to the assertion that if each S_α is not the empty set or null set, then
$$\underset{\alpha}{\times} S_\alpha$$
is not the null set. Or equivalently, if $\times S_\alpha$ is empty, then at least one S_α must be empty also.

Many of the variants of the axiom of choice involve order. We begin with a basic concept. A set S is *partially ordered*, if there exists, for some pairs of elements, a and b, in S, a relation $a \leq b$ (read a less than or equal to b), subject to the rules

1) if $a \leq b$ and $b \leq c$ (all in S) then $a \leq c$.
2) $a \leq a$. If $a \leq b$ and $b \leq a$, then $a = b$.

There are many beautiful and natural examples. Let S, for example, be a non-empty set. Let T be $P(S)$, the set of all subsets of S. If $A, B \in T$, that is A and B are subsets of S, we say $A \leq B$, if $A \subseteq B$, i.e. A is a subset of B. One easily checks that this is a partial order. It is quite possible that neither $A \leq B$ nor $B \leq A$, as the following picture would show.

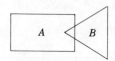

However, we always do have $A \cap B \leq A$, for any A and B.

Here are two more refined concepts. A partially ordered set is called a *directed set*, if in addition it enjoys the property that for every $x, y \in S$, we can find a $z \in S$ with $x \leq z, y \leq z$. A *partially ordered* set is called *ordered* (or sometimes linearly ordered), if for every $x \in S$, $y \in S$, we have either $x \leq y$ or $y \leq x$.

The readers should check that an ordered set is a directed set, and a directed set is of course a partially ordered set. The real numbers, with the usual notion of order, is clearly an ordered set. A final concept is that of a *well-ordered set*. An ordered set is *well-ordered* if every non-empty subset has a least element, that is an element which is less than or equal to every element of that subset. For example, the positive whole numbers or integers are well-ordered. On the other hand, the set of real numbers (decimals) with the usual order is *not* well-ordered. Indeed, the set of strictly positive real numbers has no smallest element. An important variant of the axiom of choice is the

Well-Ordering Principle

If S is any set, then there is some order relation defined on S, which makes S a well-ordered set. That is, any set has a suitable order for which every subset has a smallest element.

A final variant of the axiom of choice, due to M. Zorn, is

Zorn's Lemma

Let S be a partially ordered set. Suppose that whenever $T \subseteq S$ is a subset which is ordered with respect to the partial ordering (that is, it happens that for any $x, y \in T$, $x \leq y$ or $y \leq x$), there is $m \in S$ so that whenever $x \in T$, $x \leq m$. (m is a "biggest element" for T).

Then S has a maximal element, i.e. there is $M \in S$ such that if $M \leq z$, $z \in S$, we have $M = z$.

The content of this statement is that if every subset which is fortunate enough to be ordered is dominated by a biggest element (sometimes called a *supremum*), then the whole partially ordered set has a maximal element (which might not be unique, and hence is not really *the* biggest element).

The axiom of choice, well-ordering principle, and Zorn's lemma are all equivalent statements. That is to say, any one implies any other. We will now give a couple of implications of this sort, referring the reader to the References at the end of the chapter for full details. Suppose we wish to show that the well-ordering principle implies the axiom of choice. Let S_α be the sets of our collection, where the collection of α's is written A. Form

$$S = \bigcup_\alpha S_\alpha$$

By the well-ordering principle, S has an order for which every subset has a least element. Each S_α is a subset of S, which is just the union of the $S_{\alpha's}$, so each S_α has a least element which we write $f(\alpha)$. This defines a function

$$f: A \to \bigcup_\alpha S_\alpha$$

which has the property that $f(\beta) \in S_\beta$ for any $\beta \in A$, as desired.

Similarly, suppose Zorn's lemma is true, and we wish to prove the axiom of choice. Using the terminology above, suppose $B \subseteq A$ is a subset of the index set, and let x_B be an element of the product

$$\underset{\alpha \in B}{\times} S_\alpha,$$

that is x_B is an element (if it exists) in the Cartesian product of those S_α for all those α's which are in B. Remember, that to prove the axiom of choice, one needs to show that

$$\underset{\alpha \in A}{\times} S_\alpha$$

is non-empty (here we take the product of all the $S_{\alpha's}$).

Let T be the set of all pairs (B, x_B). We note that if $C \subseteq B$, there is a function

$$g_B{}^C: \underset{\alpha \in B}{\times} S_\alpha \to \underset{\alpha \in C}{\times} S_\alpha$$

which is defined by taking those functions $f: B \to \bigcup_{\alpha \in B} S_\alpha$ and considering them as having a smaller domain of definition than B, namely C. In fact, the functions or maps P_1 and P_2 defined earlier are just special cases of this idea. We define $(C, x_C) \leq (B, x_B)$, if $C \subseteq B$ and $g_B{}^C(x_B) = x_C$. One easily checks that this is a partial ordering on T.

If T' is a linearly ordered subset of T, we define a supremum of T' by taking the union of all B's, for which $(B, x_B) \in T'$; we easily define a "biggest" x_B by noting that on any α in the union of these B's, the value of each x_B is fixed, so that a supremum x_B is well-defined.

By Zorn's lemma, T has a maximal element which we write (M, x_m), $M \subseteq A$. I claim that $M = A$, in which case the element x_M will belong to

$$\underset{\alpha}{\times} S_\alpha$$

proving the axiom of choice. Suppose $\alpha_1 \in A$, $\alpha_1 \notin M$ (that is assume M is smaller than A). x_M is a function

$$x_M: M \to \bigcup_{\alpha \in M} S_\alpha$$

Let $m_1 \in S_{\alpha_1}$. Define $M_1 = M \cup \{\alpha_1\}$, that is the set consisting of all the elements of M plus, in addition, the element α_1. Define

$$x_{M_1}: M_1 \to \bigcup_{\alpha \in M_1} S_\alpha$$

by $x_{M_1}(\alpha) = x_M(\alpha)$ if $\alpha \in M$, and $x_{M_1}(\alpha_1) = m_1$ in the single other case. It is easy to check that $(M, x_M) \leq (M_1, x_{M_1})$ which contradicts the fact that M is maximal. Therefore, the assumption that M was smaller than A was false, and we have completed the proof.

This completes the review of set theory. We note again that one can actually show that these assertions are equivalent. Details may be found in the References (*General Topology* by J. Kelley, Ch. 0, is particularly good on these ideas).

As a final introductory topic, we speak briefly of the theory of categories, which is a young subject which is playing an increasing role of unification in mathematics. A category is a general concept which is designed to encompass any structure which has any common occurrence in modern mathematics. The idea is based on the belief that a mathematical system is defined by making precise the objects of that system (sets, groups, vector spaces, etc.) and the mappings which one wants to allow between the objects of the system. For example, the category of sets, which we have been considering earlier in this chapter, is made up of sets and functions (i.e. single valued functions from sets to sets). The category of vector spaces is made up of vector spaces and as maps, one traditionally takes the linear transformations from one vector space to another. In algebra, the category of groups has as objects the groups and as maps or functions, the homomorphisms.

The formal definition of a category is the following:

A category $\mathcal{C}$ consists of a collection of objects $A, B, \cdots$ and maps between these objects. No restriction is placed on the objects, but the maps are required to satisfy the following two axioms:

C1) For any object A there is a map $1_A: A \to A$ (which is the identity map, such as $f(x) = x$), so that if $B \xrightarrow{f} A$ and $A \xrightarrow{g} C$ are maps in the category, the composite maps (which are all assumed to be defined) satisfy

$$g \cdot 1_A = g \quad \text{and} \quad 1_A \cdot f = f$$

C2) If $A \xrightarrow{f} B$, $B \xrightarrow{g} C$, and $C \xrightarrow{h} D$ are maps in the category, we have $(h \cdot g) \cdot f = h \cdot (g \cdot f)$.

One easily checks that the examples above are all in fact categories. Frequently categories are distinguished by taking smaller and smaller classes of maps. For example, the category of sets and the category of groups have essentially the same objects because any set can be made into a group for a suitable multiplication. But the notion of homomorphism is a severe restriction on which maps are allowed.

One can define a map from one category to another as a pair of functions (usually written with the same letter) which takes objects to objects and maps to maps. Such a map $F: \mathcal{C}_1 \to \mathcal{C}_2$ from category $\mathcal{C}_1$ to category $\mathcal{C}_2$ should satisfy the rules

$$F1) \ F(1_A) = 1_{F(A)}$$
$$F2) \ F(g \cdot f) = F(g) \cdot F(f).$$

This sort of map, from one category to another, is called a *functor*. One can even define maps between functors, etc., but we have to refer to the literature, such as B. Mitchell, *Theory of Categories*, for this.

The bulk of this text consists in the study of the category whose objects are topological spaces and whose maps are the continuous functions. The construction of the fundamental group (Chapter 8) is an important example of a functor.

REFERENCES

Bourbaki, N.: *Éléments de Mathematiques*, Paris: Hermann et Cie. (Many volumes. See, in particular, the set theory and general topology. This is a *hard* reference. *Note:* Some volumes have been translated into English and published by Addison-Wesley Publishing Co., Inc., Reading, Mass.)

Halmos, P.: *Naive Set Theory*, New York: Van Nostrand Reinhold Co., 1960.

Kelley, J.: *General Topology*, New York: Van Nostrand Reinhold Co., 1955. (This text is also good for set theory.)

MacLane, S.: *Categories for the Working Mathematician*, New York: Springer-Verlag, 1971.

Mitchell, B.: *Theory of Categories*, New York: Academic Press, Inc., 1965. (This is an advanced and comprehensive text.)

A substantially more advanced logic and set theory text is:

Cohen, P. J.: *Set Theory and the Continuum Hypothesis*, Menlo Park, Calif.: W. A. Benjamin Inc., 1966.

CHAPTER 2

Metric Spaces

We propose to give a detailed study of metric spaces, which is a large and important class of topological spaces. To motivate this, we shall begin with a review of the real numbers and Euclidean spaces (the Cartesian plane, 3-dimensional space, etc.).

The real number system is universal in mathematics. One may construct it rigorously from the integers (see E. Landau, in the Bibliography). But as our approach is review, we will describe it axiomatically. Recall that a *field F* is a set on which we have an operation $+$ (an addition) and $\cdot$ (a multiplication) so that

A1) For $a, b \in F$ (that is, $a \in F$ and $b \in F$), $a + b \in F$ is defined.
A2) $(a + b) + c = a + (b + c)$ for any $a, b, c \in F$.
A3) There is $0 \in F$ so that $a + 0 = a$, for all $a \in F$.
A4) For each $a \in F$, there is $-a \in F$, with $a + (-a) = 0$.
A5) $a + b = b + a$ for all $a, b \in F$.
M1) $a \cdot b \in F$ is defined, when $a, b \in F$.
M2) $(a \cdot b) \cdot c = a \cdot (b \cdot c)$ for all $a, b, c \in F$.
M3) There is $1 \in F$, $1 \neq 0$, with $a \cdot 1 = a$, all $a \in F$.
M4) If $a \neq 0$, there is a^{-1}, so that $a \cdot a^{-1} = 1$.
M5) For all $a, b \in F$, $a \cdot b = b \cdot a$.
D) For all $a, b, c \in F$

$$a \cdot (b + c) = a \cdot b + a \cdot c$$

The A1–A5 assure that F is an abelian group under addition. M1–M5 say that the non-zero elements form an Abelian group under multiplication. D is the distributive law.

The rational numbers, real numbers, complex numbers are all fields. The set with 2 elements 0 and 1 may be easily made into a field, with suitable choice of addition and multiplication.

A field is *ordered* if there is an order defined $a \leq b$ satisfying, for all

15

$a, b, \cdots \in F$,
1) $a \leq a$.
2) If $a \leq b, b \leq c$, then $a \leq c$.
3) If $a \leq b$ and $b \leq a$, then $a = b$.
4) If $a \leq b$, $a + c \leq b + c$.
5) If $a \leq b$ and $c \geq 0$, then $a \cdot c \leq b \cdot c$.

In addition, we have the Archimedean order property,

6) If $a \leq b, a > 0$ there is an integer n with $b \leq n \cdot a$.

Finally, we have the supremum or least upper bound property for a field F.

S) If $S \subseteq F$ is a non-empty subset, which is bounded in the sense that there is a $k \in F$, with the property that if $x \in S$, $x \leq k$, then there is a smallest such k with this property.

The real numbers is the unique ordered field (i.e. satisfying A1–A5, M1–M5, D, 1–6) for which axiom S holds. The construction may be found in Landau (and other books). The uniqueness is standard in various texts. The development of the real numbers in a rigorous way is traditionally treated, at this time, in a beginning course in real variables or mathematical analysis (for example, R. Bartle, in the Bibliography). A student who is not well into the details of the real numbers should study the appropriate books in the Bibliography thoroughly before proceeding here.

We wish to study first the Euclidean spaces, which are the higher dimensional generalizations of the real numbers.

Definition 2.1

By Euclidean n-space, we mean the set of (ordered) n-tuples of real numbers, that is to say, the set of all sequences of length n, $(x_1, \cdots, x_n)$, where each x_i is a real number. We write it as E^n.

Other locutions are n-space, real n-space, etc. Observe that the real numbers consist precisely of Euclidean 1-space, and that we shall use the locution "plane" freely for Euclidean 2-space.

Traditionally, Euclidean n-space is regarded as having considerably more structure than that of just a set, and we now describe this structure.

Definition 2.2

I. Two elements of Euclidean n-space may be added by the definition

$$(x_1, \cdots, x_n) + (y_1, \cdots, y_n) = (x_1 + y_1, \cdots, x_n + y_n)$$

II. If $\alpha \in R$ is a real number (or equivalently $\alpha \in E^1$), we define a scalar product

$$\alpha \cdot (x_1, \cdots, x_n) = (\alpha x_1, \cdots, \alpha x_n)$$

where αx_1, etc., refers to the product of real numbers α and x_1.

Metric Spaces 17

This addition and scalar product satisfy various laws. Suppose we write for short.

$$X = (x_1, \cdots, x_n), \quad Y = (y_1, \cdots, y_n)$$
$$Z = (z_1, \cdots, z_n) \quad \theta = (0, \cdots, 0)$$

so that $X, Y, Z, \theta \in E^n$. Suppose $\alpha, \beta, \ldots$ are real numbers. Then you may (and should) easily check the following proposition.

Proposition 2.1

a) $(X + Y) + Z = X + (Y + Z)$
b) $X + \theta = X$
c) $X + Y = Y + X$
d) $X + (-1)X = \theta$
e) $(\alpha + \beta) \cdot X = \alpha \cdot X + \beta \cdot X$
f) $\alpha \cdot (X + Y) = \alpha \cdot X + \alpha \cdot Y$
g) $(\alpha \cdot \beta) \cdot X = \alpha \cdot (\beta \cdot X)$
h) $1 \cdot X = X$.

For example, we have as a proof of e),

$$(\alpha + \beta) \cdot X = ((\alpha + \beta)x_1, (\alpha + \beta)x_2, \cdots, (\alpha + \beta)x_n)$$
$$= (\alpha x_1 + \beta x_1, \alpha x_2 + \beta x_2, \cdots, \alpha x_n + \beta x_n)$$
$$= (\alpha x_1, \cdots, \alpha x_n) + (\beta x_1, \cdots, \beta x_n) = \alpha \cdot X + \beta \cdot X.$$

The first equality follows from the Definition 2.2 (II), the second from the distributive law for real numbers, and the third from the definition of addition ((I) above); all the other proofs follow directly from the definitions.

Remarks. 1) If you know linear algebra, you will recognize that Proposition 2.1 says that E^n is a vector space over the real numbers.

2) In the case of the plane, our definition of addition coincides exactly with the traditional definition of adding vectors (the diagonal of the parallelogram).

3) E^n has no natural definition of order except in the case of $E^1 = \mathbb{R}$.

4) $\theta = (0, 0, \cdots, 0)$ is usually referred to as the origin.

In addition, E^n has a particularly simple and beautiful notion of distance, whose definition is based upon a generalization of the familiar Pythagorean theorem.

Definition 2.3

If $X = (x_1, \cdots, x_n)$ and $Y = (y_1, \cdots, y_n)$ are elements of E^n, we define the distance from X to Y by

$$d(X, Y) = \sqrt{(x_1 - y_1)^2 + (x_2 - y_2)^2 + \cdots + (x_n - y_n)^2}.$$

18   Topology

We note that $d(X, X) = 0$, while if $X \neq Y$, then $d(X, Y)$ must be positive. Also it is clear that $d(X, Y) = d(Y, X)$. To get deeper into the notion of distance, we introduce the inner product.

Definition 2.4

$$\langle X, Y \rangle = x_1 y_1 + x_2 y_2 + \cdots + x_n y_n.$$

This concept produces, for each pair of elements in E^n, a real number. One may easily check, from the definitions, the following facts.

Proposition 2.2

a) $\langle X, Y \rangle = \langle Y, X \rangle$
b) $\langle X, \theta \rangle = 0$
c) $\langle X, Y + Z \rangle = \langle X, Y \rangle + \langle X, Z \rangle$
d) $\alpha \langle X, Y \rangle = \langle \alpha X, Y \rangle = \langle X, \alpha Y \rangle$.

For example, if $(1, 2, 3)$ and $(-2, 5, 0)$ are in E^3, then

$$\langle (1, 2, 3), (-2, 5, 0) \rangle = 8$$

$\langle X, Y \rangle = 0$ means, in a general sense, that X is perpendicular to Y.

A more penetrating fact is the following inequality, due to Cauchy and Schwartz.

Proposition 2.3

For any $X, Y \in E^n$,

$$\langle X, Y \rangle \leq (\sqrt{\langle X, X \rangle})\, (\sqrt{\langle Y, Y \rangle}).$$

This makes sense, because clearly $\langle X, X \rangle = (d(X, 0))^2 \geq 0$.

Proof. Let α be any real number. Then, we know that

$$0 \leq \langle (X + \alpha \cdot Y), (X + \alpha \cdot Y) \rangle$$
$$= \langle X, (X + \alpha Y) \rangle + \langle \alpha Y, (X + \alpha Y) \rangle$$
$$= \langle X, X \rangle + \alpha \langle X, Y \rangle + \alpha \langle Y, X \rangle + \alpha^2 \langle Y, Y \rangle$$
$$= \langle X, X \rangle + 2\alpha \langle X, Y \rangle + \alpha^2 \langle Y, Y \rangle.$$

All these steps follow easily from the basic facts in Proposition 2.2. We thus have a quadratic expression

$$\alpha^2 \langle Y, Y \rangle + 2\alpha \langle X, Y \rangle + \langle X, X \rangle$$

which we know cannot be negative. It represents a parabola, in the variable α, with constants $\langle X, X \rangle$, $\langle X, Y \rangle$, and $\langle Y, Y \rangle$, which must never cross the axis.

Or in other words, the equation
$$\langle Y, Y\rangle\alpha^2 + 2\langle X, Y\rangle\alpha + \langle X, X\rangle = 0$$
can have either one root (parabola just touches the axis) or no roots (parabola lies strictly above the axis). It can never have two distinct real roots. From the theory of quadratic equations, we then know that
$$(2\cdot\langle X, Y\rangle)^2 - 4\langle Y, Y\rangle\langle X, X\rangle \leq 0,$$
it being less than zero when there are no real roots and equal to zero when there is precisely one root. The desired inequality follows immediately.

This proposition has an important corollary, sometimes called the Minkowski inequality.

Proposition 2.4

We write $\|X\| = d(X, \theta)$. It is often called the *length* of X. We have
$$\|X + Y\| \leq \|X\| + \|Y\|.$$

Proof. Using the facts above, we calculate
$$\|(X + Y)\|^2 = d(X + Y, \theta)^2$$
$$= \langle X + Y, X + Y\rangle = \langle X, X\rangle + 2\langle X, Y\rangle + \langle Y, Y\rangle$$
$$= \|X\|^2 + 2\langle X, Y\rangle + \|Y\|^2.$$

Now, we shall replace $\langle X, Y\rangle$ by the bigger $\|X\|\cdot\|Y\|$ according to Proposition 2.3. We then get
$$\|X + Y\|^2 \leq \|X\|^2 + 2\|X\|\|Y\| + \|Y\|^2 = (\|X\| + \|Y\|)^2$$
Taking the square root of both sides, which is all right since everything here is positive, we have the claimed assertion.

Corollary

$$d(X, Y) \leq d(X, Z) + d(Z, Y) \quad \text{for all } X, Y, Z \in E^n.$$

Proof. We easily check that
$$d(X, Y) = \|X - Y\|$$
from the definitions. But then, Proposition 2.4 gives
$$\|X - Y\| = \|(X - Z) + (Z - Y)\| \leq \|X - Z\| + \|Z - Y\|,$$
or $d(X, Y) \leq d(X, Z) + d(Z, Y)$ as required.

We summarize our study of the distance function by the following three, very basic properties, which hold for all $X, Y, Z \in E^n$.

A) $d(X, Y) \geq 0$. $d(X, Y) = 0$ precisely when $X = Y$, that is for each i, $x_i = y_i$.
B) $d(X, Y) = d(Y, X)$.
C) $d(X, Y) \leq d(X, Z) + d(Z, Y)$.

The last assertion generalizes the basic fact, from Euclidean geometry that the sum of two sides of a triangle is bigger than the third.

Problems

1. A closed interval in the real numbers is a set of the form

$$\{x \mid a \leq x \leq b\}$$

   Show that if two closed intervals meet, their intersection is again a closed interval.

2. Show that if a and b are two distinct real numbers, there are infinitely many real numbers between them.

3. Prove that if ϵ is a positive real number, there is an integer n, with

$$0 < \frac{1}{n} < \epsilon.$$

   (*Hint:* Use property 6 of the order relation.)

4. Writing $\epsilon = (b - a)$, show, by using Exercise 3, that there are whole numbers m and n such that

$$a \leq m/n \leq b.$$

   Conclude that between any two real numbers, we may find a rational number.

5. If $X \in E^2$ is any non-zero element, how many Y's are there in E^2 satisfying
   a) $\| Y \| = 1$
   b) $\langle X, Y \rangle = 0$?

6. Let $X, Y \in E^2$. Let A be the area of a parallelogram with sides of length $\| X \|$ and $\| Y \|$, which meet in an angle equal to that angle of the lines joining X to θ and Y to θ.
   Show that the $A \leq \| X \| \cdot \| Y \|$.
   Show that this inequality is in fact an equality, if and only if $\langle X, Y \rangle = 0$.

7. Using trigonometry, show that if α is the angle between the lines joining X to θ and Y to θ, in E^2, X and Y both not θ,

$$\cos \alpha = \frac{\langle X, Y \rangle}{\| X \| \cdot \| Y \|}.$$

8. Let $\delta(X, Y)$ be a function, in two variables, which assigns to each pair of elements in E^n (or more generally any set) a real number.

Suppose for all X, Y, Z in E^n
A) $\delta(X, Y) \geq 0$
   $\delta(X, Y) = 0$ if and only if $X = Y$.
C^1) $\delta(X, Y) \leq \delta(X, Z) + \delta(Y, Z)$.
Deduce that
B) $\delta(X, Y) = \delta(Y, X)$
and thus
C) $\delta(X, Y) \leq \delta(X, Z) + \delta(Z, Y)$.

9. If $n < m$, one may find various subsets of E^m which are identical to E^n in the sense that there is a 1-1 correspondence between the points of E^n and the subset of E^m under which distance is preserved (two points in E^n, at distance d, are carried over to two points in the subset of E^m at distance d). For example, there is a 1-1 correspondence of E^2 and the subset

$$\{(x_1, x_2, x_3) \mid (x_1, x_2, x_3) \in E^3, x_3 = 5\}$$

which preserves distance.

Find two subsets of E^4 which are identical to E^2, and which intersect in a single point.

We wish next to investigate some special properties of the spaces E^n, with the purpose of showing that many of the properties of the plane familiar from Euclidean geometry, also hold in this more general setting.

Definition 2.5

A line in E^n may be defined as a set of points $(x_1, \cdots, x_n)$ subject to the conditions

$$x_1 = a_1 t + b_1$$
$$\vdots$$
$$x_n = a_n t + b_n$$

where $a_1, \cdots, a_n, b_1, \cdots, b_n$ are constants and t is a variable real number.

If we use the shorthand

$$X = (x_1, \cdots, x_n)$$
$$A = (a_1, \cdots, a_n)$$
$$B = (b_1, \cdots, b_n),$$

then we can write

$$X = t \cdot A + B.$$

This generalizes the traditional parametric definition of a line in a plane. Similarly, we may define a line segment, or an interval in a line, by the same equation, and where t varies over an interval $t_0 \leq t \leq t_1$. We now show that any two points in E^n may be joined by a line segment.

Proposition 2.5

If $P = (p_1, \cdots, p_n)$ and $Q = (q_1, \cdots, q_n)$ are two points or elements in E^n, there is a line segment joining P and Q in E^n.

Proof. Writing $-Q$ for $(-1) \cdot Q$, define

$$X = t \cdot (P - Q) + Q$$

where $0 \leq t \leq 1$. Then, when $t = 0$, $X = Q$ and when $t = 1$, $X = P$, so that this line segment joins Q and P as desired.

A set S in E^n is called *convex*, if, whenever P and Q belong to S, the entire line segment from P to Q (or Q to P) belongs to S. Proposition 2.5 says, in this language, that the entire space E^n is convex. Various important subsets of E^n, such as the closed and open balls defined below, are in fact also convex. Such sets play an important role in various kinds of geometry.

Students familiar with analysis will know the traditional definition of closed and open intervals of real numbers

$$[a, b] = \{x \mid a \leq x \leq b\}$$
$$\langle a, b \rangle = \{x \mid a < x < b\}.$$

We may easily rewrite the definition of a closed interval as

$$[a, b] = \left\{ x \mid \left| x - \frac{a+b}{2} \right| \leq \frac{b-a}{2} \right\},$$

because $(a + b)/2$ is the center of the segment from a to b and $b - a$ is the diameter of the segment. A similar definition holds for an open interval replacing $\leq$ by $<$. The following definitions of balls in E^n are obvious generalizations of these intervals in E^1 or $\mathbb{R}$.

Definition 2.6

We define an open n-ball in E^n by

$$B_R{}^n(C) = \{X \mid X \in E^n, d(X, C) < R\}$$

and the closed n-ball by

$$\bar{B}_R{}^n(C) = \{X \mid X \in E^n, d(X, C) \leq R\}$$

C is referred to as the center of the ball and R the radius.

Definition 2.7

An $(n - 1)$-sphere is defined as

$$\bar{B}_R{}^n(C) - B_R{}^n(C) = S_R{}^{n-1}(C)$$

where $R > 0$ is the radius and C the center. That is to say, an $(n-1)$-sphere is the set of the form

$$\{X \mid X \in E^n, d(X, C) = R\}$$

(where $C \in E^n$ is a (fixed) center and the constant $R > 0$ is the radius).

Note that the balls $B_R(C)$ and $\bar{B}_R(C)$ are essentially n-dimensional things, while the sphere $\bar{B}_R{}^n(C) - B_R{}^n(C)$ is an $(n-1)$-dimensional object.

In E^2, we represent $\bar{B}_1{}^n(\theta)$, $B_1{}^n(\theta)$, and $S_1{}^{n-1}(\theta)$ pictorially as follows:

The dotted circle means that the points, whose distance from the origin is precisely 1, are to be omitted.

Problems

1. Write the equations which describe the coordinate axes in E^3.
2. Let $X \in B_R{}^n(\mathfrak{C}) \subset E^n$, where $R > 0$. Let $d(X, \mathfrak{C}) = d < R$. Find the shortest distance from X to a point on $S_R{}^{n-1}(\mathfrak{C})$. (*Hint:* Use the fact that the distance between two points is the length of the line between them and that if A, B, C lie in a row on a line, the distance from A to B plus that from B to C equals that from A to C. Draw a picture.)
3. Prove that the sets $B_R{}^n(\mathfrak{C})$ and $\bar{B}_R{}^n(\mathfrak{C})$ are convex. That is if A and B are points in one of these sets, then every point on the line segment from A to B lies in that set.
4. Prove that $S_R{}^{n-1}(\mathfrak{C})$ is *not* convex.
5. Locate $(n+1)$-points in E^n so that the distance between any two is 1. This gives a generalization of the triangle and tetrahedron. The figure made up by this configuration is called an n-simplex.

We may now begin the study of metric spaces, which are on the one hand a natural generalization of Euclidean spaces, and on the other hand a very important class of general topological spaces. Every reasonable non-pathological space in topology (the general definition in the next chapter) will turn out to be a metric space. On the other hand, the developments in analysis in the early 20th century showed that there was a need to study a more general class of spaces than merely Euclidean spaces.

The definition of a metric space is really very simple, consisting of a set and a meaningful notion of distance on that set (that is, the properties

24   Topology

which are called $A)$, $B)$ and $C)$ earlier, for the distance function on E^n). We first give the definition, and then consider a wide variety of important examples.

Definition 2.8

A set S is a metric space, if, for every pair of points $X, Y \in S$, there is a well-defined function $d(X, Y)$, whose values are real numbers, so that
A) $d(X, Y) \geq 0$. $d(X, Y) = 0$, if and only if $X = Y$.
B) $d(X, Y) = d(Y, X)$
C) $d(X, Y) \leq d(X, Z) + d(Z, Y)$
for any $Z \in S$.

Such a distance function d is also called a *metric*.

Examples. 1) The real numbers $\mathbb{R}$. The distance, defined by $d(a, b) = |b - a|$, clearly satisfies our three conditions.

2) Euclidean space E^n with the distance defined earlier in this chapter.

3) S is a finite set of points, say $a, b, c, \ldots$ with the distance between any point and itself equal to zero and the distance between two distinct points equal to 1.

4) Let S be a metric space, with distance function $d(X, Y)$. Define a new distance function $d'(X, Y)$ where

$$d'(X, Y) = \alpha \cdot d(X, Y), \alpha \in \mathbb{R}, \alpha > 0.$$

Then S is a metric space and if $\alpha \neq 1$, d' is a new distance function.

The following examples are of paramount importance in analysis.

5) We take, as our set, the set of continuous real functions defined on $[a, b] = \{x \mid a \leq x \leq b\}$. Denote this $\mathcal{C}[a, b]$. Let $f, g \in \mathcal{C}[a, b]$. Define

$$d(f, g) = \underset{a \leq x \leq b}{\operatorname{maximum}} \left(|f(x) - g(x)| \right)$$

That is the maximum of $|f(x) - g(x)|$ for all x in $[a, b]$.

This makes sense because a continuous function, here $|f(x) - g(x)|$, defined on a closed interval, here $[a, b]$, is well-known to have a maximum. (In fact, we shall prove a more general assertion, which will imply this, in Chapter 4.)

To prove A), B), and C), note that $d(f, g)$ is clearly non-negative, and if $d(f, g) = 0$, then $|f(x) - g(x)| = 0$ for all x. Then $f(x) = g(x)$ for all x.

Since $|f(x) - g(x)| = |g(x) - f(x)|$, clearly B) holds, and we need only prove C) to complete the proof that $\mathcal{C}[a, b]$ is a metric space.

We note that

$$|f(x) - g(x)| = |f(x) - h(x) + h(x) - g(x)|$$
$$\leq |f(x) - h(x)| + |h(x) - g(x)|.$$

Let x_0 be such that $|f(x) - g(x)|$ is maximum. Because the maximum of $|f(x) - h(x)|$ is at least as big as $|f(x_0) - h(x_0)|$, (the maximum is always at least as big as any specific value) and the maximum of $|h(x) - g(x)|$ is at least as big as $|h(x_0) - g(x_0)|$, we get $\max(|f(x) - g(x)|) = |f(x_0) - g(x_0)| \leq |f(x_0) - h(x_0)| + |h(x_0) - g(x_0)| \leq \max(|f(x) - h(x)|) + \max(|h(x) - g(x)|)$, where max refers to the maximum value on $[a, b]$. But then by definition

$$d(f, g) \leq d(f, h) + d(h, g)$$

as desired.

6) (Hilbert Space) Let H be the set of all sequences of real numbers $\{a_1, a_2, \cdots\}$ subject to the condition that

$$\sum_{i=1}^{\infty} a_i^2$$

converges. It is clear, from our Proposition 2.3, that

$$\left(\sum_{i=1}^{n} |a_i b_i|\right)^2 \leq \sum_{i=1}^{n} a_i^2 \sum_{i=1}^{n} b_i^2$$

where $\{b_1, b_2, \cdots\}$ is another such sequence. It follows that if the two sequences in question belong to H, then

$$\sum_{i=1}^{\infty} a_i b_i = \langle (a_1, a_2, \cdots), (b_1, b_2, \cdots) \rangle$$

converges (because an absolutely convergent series is convergent and a positive series, whose partial sums are bounded, is convergent.)

We may then make the following definition.

If $A = (a_1, a_2, \cdots)$ and $B = (b_1, b_2, \cdots)$ belong to H, then we put

$$d(A, B)^2 = \sum_{i=1}^{\infty} (a_i - b_i)^2$$

(The earlier remarks show that the series in question here actually converges.)

The properties A) and B) are clear, while C) follows immediately from an easy generalization of Proposition 2.4. (See various books on analysis, in the Bibliography, etc.)

To show that the collection of metric spaces is indeed very large, we prove now two basic, easy propositions.

Proposition 2.6

If S is a metric space, with distance function d, and T is a subset of S, then T is also a metric space using the same notion of distance.

26   Topology

Proof. It is clear that if $x, y \in T \subseteq S$, $d(x, y)$ is defined. d satisfies all properties A), B), and C) for all elements of the smaller T.

We note that this idea will figure in Chapter 5, when we prove that every space satisfying certain reasonable hypotheses is actually a metric space.

Proposition 2.7

If S_1 and S_2 are metric spaces, with d_1 and d_2 being the distance functions, respectively, then $S_1 \times S_2$ is a metric space with distance function

$$d((x_1, x_2), (y_1, y_2)) = \sqrt{(d_1(x_1, y_1))^2 + (d_2(x_2, y_2))^2}.$$

Note that this definition of d is given in terms of the known metrics d_1 and d_2 by a formula which is modelled after the traditional Pythagorean theorem.

Proof. d is clearly non-negative, and if $x_1 = y_1$ and $x_2 = y_2$, then d is 0. Suppose

$$d((x_1, x_2), (y_1, y_2)) = 0,$$

then $d_1(x_1, y_1) = 0$ and $d_2(x_2, y_2) = 0$, because otherwise the sum of their squares would be positive. But because d_1 and d_2 are metrics, $x_1 = y_1$ and $x_2 = y_2$, so that $(x_1, x_2) = (y_1, y_2)$ in $S_1 \times S_2$. This proves A). B) is obvious, and we now prove C).

If (z_1, z_2) is a third point or element in $S_1 \times S_2$, we write the inequality C) with respect to this third point for the two metrics, squaring both sides. We get

$$d_1(x_1, y_1)^2 \leq d_1(x_1, z_1)^2 + d_1(z_1, y_1)^2 + 2d_1(x_1, z_1)d_1(z_1, y_1)$$

and

$$d_2(x_2, y_2)^2 \leq d_2(x_2, z_2)^2 + d_2(z_2, y_2)^2 + 2d_2(x_2, z_2)d_2(z_2, y_2);$$

adding the two inequalities together and using the definition of d as the metric on $S_1 \times S_2$, we get

$$d((x_1, x_2), (y_1, y_2))^2 \leq d((x_1, x_2), (z_1, z_2))^2 + d((z_1, z_2), (y_1, y_2))^2$$
$$+ 2[d_1(x_1, z_1)d_1(z_1, y_1) + d_2(x_2, z_2)d_2(z_2, y_2)].$$

The proof will be done if we can show that the square brackets are less than or equal to

$$2\sqrt{d_1(x_1, z_1)^2 + d_2(x_2, z_2)^2}\sqrt{d_1(z_1, y_1)^2 + d_2(z_2, y_2)^2}$$

because then the inequality C) for the metric d will fall out upon taking the square root (easy check).

To obtain this final step, note that for any numbers A, B, C, D, all posi-

tive,
$$0 \leq (B \cdot C - A \cdot D)^2 = B^2C^2 + A^2D^2 - 2ABCD$$
or
$$2ABCD \leq B^2C^2 + A^2D^2$$
or
$$A^2B^2 + 2ABCD + C^2D^2 \leq A^2B^2 + B^2C^2 + A^2D^2 + C^2D^2$$
or
$$(AB + CD) \leq \sqrt{A^2 + C^2} \cdot \sqrt{B^2 + D^2}.$$

Let $A = d_1(x_1, z_1)$, $B = d_1(z_1, y_1)$, $C = d_2(x_2, z_2)$, $D = d_2(z_2, y_2)$ and the desired step is complete.

In general, the proof is not nearly as difficult as the complicated terminology would make it appear to be, each step being a matter of very elementary algebra.

These two propositions enable us to generate a very large family of examples of metric spaces. We see at once that $B_R^n(\mathcal{C})$, $\bar{B}_R^n(\mathcal{C})$, $S_R^{n-1}(\mathcal{C})$ are all metric spaces as well as their various Cartesian products.

We may now define continuous mappings, of metric spaces, and introduce the basic concept of open set, which will then be studied in greater generality in the next chapter.

Definition 2.9

Let $f: S_1 \to S_2$ be a function from the metric space S_1 (with distance function d_1) to the metric space S_2 (distance function d_2). f is, of course, presumed to be single valued, i.e. for each $x \in S_1$, there is precisely one $f(x) \in S_2$.

We define:

A) f is *continuous at* $x_0 \in S_1$, if given any $\epsilon > 0$, there is a $\delta > 0$ such that whenever $d_1(y, x_0) < \delta$, then $d_2(f(y), f(x_0)) < \epsilon$.

In other words, for any $\epsilon > 0$ (which is a measure of accuracy and we mean especially the small ϵ's), we can find a $\delta > 0$ such that all $y \in S$, which are closer to x_0 than δ, are mapped into points of S_2, which are nearer $f(x_0)$ than our measure of accuracy. In terms of pictures, and rather unrigorously, we could say that points of S_1 which are near x are not ripped away from $f(x_0)$ by the map f.

B) f is *continuous*, if it is continuous at all $x_0 \in S_1$.

Remarks. This definition clearly generalizes the definition of continuous function in calculus, where the function goes from $\mathbb{R}$ to $\mathbb{R}$ and where $d(x, y) = |x - y|$.

Many functions are visibly continuous. For example, the constant functions and the identity functions, defined (respectively) as

$$C: S_1 \to S_2 \quad \text{by} \quad C(x) = z_0, \quad \text{for some fixed} \quad z_0 \in S_2$$

$$1: S_1 \to S_1 \quad \text{by} \quad 1(x) = x \quad \text{for all} \quad x \in S_1.$$

If $f_1: S \to T_1$ and $f_2: S \to T_2$, both continuous, then

$$f_1 \times f_2: S \to T_1 \times T_2 \quad \text{defined by} \quad (f_1 \times f_2)(x) = (f_1(x), f_2(x))$$

is continuous. Let $\epsilon > 0$, $x_0 \in S$. Suppose we select $\delta_1 > 0$ and $\delta_2 > 0$ so that if d, d_1, d_2 are the respective distances in S, T_1, and T_2, then when $d(y, x_0) < \delta_1$, $d_1(f_1(y), f_1(x_0)) < \epsilon/2$ and when $d(y, x_0) < \delta_2$, $d_2(f_2(y), f_2(x_0)) < \epsilon/2$. Then if $d(y, x_0) <$ minimum (δ_1, δ_2), we have

$$\sqrt{d_1(f_1(y), f_1(x_0))^2 + d_2(f_2(y), f_2(x_0))^2} \leq \sqrt{\frac{\epsilon^2}{4} + \frac{\epsilon^2}{4}} = \sqrt{\frac{\epsilon^2}{2}} = \frac{\epsilon}{\sqrt{2}} < \epsilon.$$

Since the square root represents the distance in $T_1 \times T_2$, we have shown that $f_1 \times f_2$ is continuous for any $x \in S$.

If $f: S \to T$ and $g: T \to U$ are two continuous functions defined for these metric spaces, namely S, T, and U, then the reader may easily check that the composite function

$$g \circ f: S \to U$$

defined by $g \circ f(x) = g(f(x))$ is continuous.

The notion of continuity brought out, in the minds of mathematicians in the early 20th century, the importance of sets with the property that if a point belongs to the set, all sufficiently nearby points belong to the set This was codified in the following important definition.

Definition 2.10

Let S be a metric space, with metric d. Suppose $V \subseteq S$ is a subset of S. V is called *open* or an *open set*, if, given $x \in V$, there is a $\delta > 0$ so that

$$B_\delta(x) \subseteq V \subseteq S$$

where $B_\delta(x) = \{y \mid y \in S, d(x, y) < \delta\}$.

In other words, a set V is open if, for each $x \in V$, there is a $\delta > 0$ such that any point $y \in S$, which is closer to x than δ, is necessarily also in V. $B_\delta(x)$ is called the open ball, of radius δ about x.

For example, an open interval of real numbers, $\langle a, b \rangle = \{x \mid x \in \mathbb{R}, a < x < b\}$ is open. For, if $x_0 \in \langle a, b \rangle$, put $\delta =$ minimum $(|x_0 - a|, |b - x_0|)$, that is the smaller of the two numbers $|x_0 - a|$ or $|b - x_0|$. Then clearly, if y is closer to x_0 than δ, y must be bigger than a and less than b, and hence $y \in \langle a, b \rangle$.

On the other hand, a single point in the real numbers is visibly not open.

As a further example, if S is any metric space, with metric d, and $B_R(x_0) = \{x \mid x \in S, d(x, x_0) < R\}$, then $B_R(x_0)$ is open. For let $x \in B_R(x_0)$. Select $\delta = R - d(x, x_0)$. Suppose $d(y, x) < \delta$. Then, we calculate, using property C for a metric,

$$d(y, x_0) \leq d(y, x) + d(x, x_0) = d(y, x) + (R - \delta) < \delta + (R - \delta) = R.$$

Hence, if $d(y, x) < \delta$, $y \in B_R(x_0)$ showing that $B_R(x_0)$ is open. This proof may be pictured in the case where S is E^2 as follows:

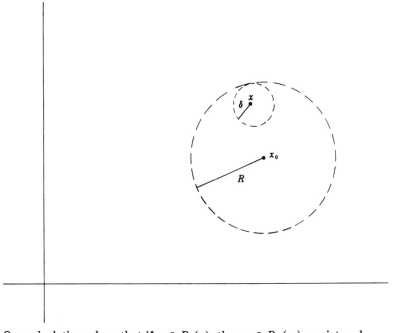

Our calculations show that if $y \in B_\delta(x)$, then $y \in B_R(x_0)$ as pictured.

Note that x_0 is always an element of $B_R(x_0)$. If S consists of only one point, it would be the only element (a rather degenerate example).

Proposition 2.8

A) Let S be a metric space and $\{V_\alpha\}$, $\alpha \in A$, be any collection (not necessarily finite) of open sets in S

Then

$$\bigcup_\alpha V_\alpha$$

is open.

B) Let $V_1, \cdots, V_n$ be a finite collection of open sets in S. Then

$$V_1 \cap V_2 \cap \cdots \cap V_n$$

is open.

Proof. Let $x \in \bigcup_\alpha V_\alpha$. Then $x \in V_\beta$ for at least one β. Since V_β is open, there is $\delta > 0$, so that if $y \in B_\delta(x)$, then $y \in V_\beta$; hence $y \in \bigcup_\alpha V_\alpha$. In other words, we have shown that if y is closer to x than δ, y is in $\bigcup_\alpha V_\alpha$, proving that this union is an open set.

To prove B), for each i, $1 \leq i \leq n$, select $\delta_i > 0$ so that if $y \in S$ is closer to $x \in V_1 \cap \cdots \cap V_n$ than δ_i, then $y \in V_i$. Let δ be the minimum of all $\delta_1, \cdots, \delta_n$. This is still a positive number, though perhaps quite small.

If y is closer to x than δ, then of course y is closer to x than each δ_i. But then, by our first remarks $y \in V_i$ for each i. This means that such a y belongs to $V_1 \cap \cdots \cap V_n$. Hence, we have shown that if y is closer to x, $x \in V_1 \cap \cdots \cap V_n$, than δ, y belongs to $V_1 \cap \cdots \cap V_n$, completing the proof.

The upshot of this proposition is that any union and any finite intersection of open sets is open. As a special case, the whole space S is open (also easy to check from the definition), because each x belongs to some open set $B_R(x)$, $R > 0$, and the union of all these is clearly S. The empty set ϕ is clearly open, because the condition which has to be satisfied never arises. That is, we should check that if $y \in \phi$, then something happens (i.e. $B_\delta(y) \subseteq \phi$). But $y \in \phi$ never happens, so that ϕ clearly satisfies the condition to be an open set.

We also note that if the metric space S consists of more than two points, then the empty set ϕ, which is a subset of S, may be written as the intersection of two non-empty open sets, i.e. if $x, y \in S$, $d(x, y) = \delta$, then

$$\phi = B_{\delta/2}(x) \cap B_{\delta/2}(y).$$

We need only check that no $z \in S$ can belong to $B_{\delta/2}(x)$ and $B_{\delta/2}(y)$ simultaneously. But if there were such a z, then

$$\delta = d(x, y) \leq d(x, z) + d(z, y) < (\delta/2) + (\delta/2) = \delta.$$

Since δ cannot be strictly less than itself, the claim is proved.

Finally, we remark that we do *not* define $\bar{B}_R(x)$ in an arbitrary metric space, although we have defined the concept for the spaces E^n. This is because of a possible confusion with the closure of a set, which will be introduced in the *following* chapter, and which will be denoted $\overline{B_R(x)}$, the bar extending over the whole set. In general, what would be a logical definition of $\bar{B}_R(x)$ is larger than the closure $\overline{B_R(x)}$ (see exercises in the next chapter), and some confusion would be possible.

Metric Spaces 31

We wish to show the intimate relationship between open sets and continuity. First we need a definition.

Definition 2.11

Let $f: S \to T$ be a map of sets. Let $V \subseteq T$. Define $f^{-1}(V) = \{x \mid x \in S, f(x) \in V\}$. In other words, $f^{-1}(V)$ is the set of points in S which f takes into V. $f^{-1}(V)$ is called the *inverse image* of V.

We now have

Theorem 2.1

Let S_1 and S_2 be metric spaces (with metrics d_1 and d_2). Suppose $f: S_1 \to S_2$ is a function or map.

Then f is continuous, if and only if whenever $V \subseteq S_2$ is an open set in S_2, then $f^{-1}(V) \subseteq S_1$ is an open set in S_1.

Proof. First, suppose that the condition that $f^{-1}(V)$ is open is true. Consider any $x \in S_1$. Let us choose for V, the ball

$$B_\epsilon(f(x)) = \{y \mid y \in S_2, d_2(f(x), y) < \epsilon\}.$$

Since $f^{-1}(B_\epsilon(f(x)))$ is open, and since $x \in f^{-1}(B_\epsilon(f(x)))$, we have some $\delta > 0$ such that

$$B_\delta(x) \subseteq f^{-1}(B_\epsilon(f(x))).$$

But this means that whenever $x_1 \in B_\delta(x)$, or equivalently, $d_1(x, x_1) < \delta$, then $f(x_1) \in B_\epsilon(f(x))$, or equivalently $d_2(f(x), f(x_1)) < \epsilon$. This is exactly the definition of continuity at x. Because x was any point of S_1, f is then a continuous function.

We must prove this converse assertion. That is, suppose f is continuous according to our Definition 2.9.

Let $V \in S_2$ be any open set, and suppose $f^{-1}(V)$ is non-empty. Of course, if $f^{-1}(V)$ were empty, which means that no point of V meets the image of f (the set of points $\{f(x) \mid x \in S_1\}$), then it would be trivially open. We wish to prove that $f^{-1}(V)$ is open. Let $x \in f^{-1}(V)$. Then $f(x) \in V$, and because V is open, there is some $\epsilon > 0$ such that

$$B_\epsilon(f(x)) \subseteq V.$$

Using the definition of continuity, there is $\delta > 0$ so that if $d_1(x, x_1) < \delta$, then $d_2(f(x), f(x_1)) < \epsilon$. But this says directly that

$$f(B_\delta(x)) \subseteq B_\epsilon(f(x)) \subseteq V$$

or intuitively that f sends points nearer to x than δ into points nearer to $f(x)$ than ϵ, and hence in V.

But if $f(B_\delta(x)) \subseteq V$, it follows at once from the definition of f^{-1} that
$$x \in B_\delta(x) \subseteq f^{-1}(V).$$
We have shown that for an $x \in f^{-1}(V)$ there is an open ball about x contained in $f^{-1}(V)$. This means that $f^{-1}(V)$ is open, finishing the proof.

This very basic Theorem 2.1 shows how the notion of open set is intimately tied up to that of a continuous function. In the next chapter, where we consider spaces which are more general and do not have metrics or distances, we shall use this idea to give a very general definition of a continuous function or map.

This completes the general theory of metric spaces. Before passing to Chapter 3, we wish, however, to stop to look at the special case of the real numbers. In a general metric space, even one as simple as the plane, the open sets can be very complicated. But in the case of the line R or E^1, one can describe the situation rather thoroughly and this is generally useful for gaining a thorough understanding of the subject.

Recall that in R, the distance function is given by
$$d(x_0, x_1) = |x_1 - x_0|.$$
We shall indicate the proof of the following characterization of the open sets in R.

Proposition 2.9

A non-empty open set in R is a disjoint union (that is a union without overlap) of open intervals, possibly including the infinite open intervals:
$$\langle -\infty, \infty \rangle = R$$
$$\langle -\infty, b \rangle = \{x \mid x \in R, x < b\}$$
$$\langle a, +\infty \rangle = \{x \mid x \in R, x > a\}.$$

Proof. The proof consists in handling various cases, each of which is fairly easy. Let O be the open set.

If every point in R is in O, $O = \langle -\infty, \infty \rangle = R$ and we are done.

Let $x \in O$. Suppose there are points b_1 and a_1, $x < b_1$, $a_1 < x$, with both a_1 and b_1 not in O, that is $a_1, b_1 \in R - O$. The set of points β with $\beta > x$, $\beta \in R - O$ is bounded from below by x. Hence it has a greatest lower bound b. The set of points $\alpha < x$, $\alpha \in R - O$, is bounded from above by x. Hence it has a least upper bound, say a. This can be drawn as follows.

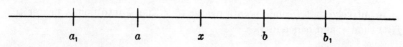

Clearly, $a < x$ for we know $a \leq x$ and, if $a = x$, then x is the least upper bound to the points less than x and not in O. This contradicts the assumption that O is open, because there must be some $\delta > 0$ so that all points nearer to x than δ are in O. But then, clearly, $x - \delta/2$ is also bigger than all points less than x and not in O, which would be impossible if x were the *least* upper bound. Similarly, one shows that $x < b$.

I claim that $\langle a, b \rangle \subseteq O$, that is all points bigger than a and less than b lie in O. This is obvious, for if there were such a point not in O, one of the claims that 1) a is the least upper bound to points less than x and not in O or 2) b is the greatest lower bound to points greater than x and not in O would be false. (Draw the picture!) One easily checks, from the definition of a and b and the definition of an open set, that a and b are not in O.

We have shown that in this case, O is the disjoint union of $\langle a, b \rangle$ and some other open set O'. If, for example, there were $a_1 < x$, not in 0, but no such b_1, then the same argument shows that O is the disjoint union of $\langle a, \infty \rangle$ and some other open set. Similarly, if there is b_1 but no a_1, O is the disjoint union if $\langle -\infty, b \rangle$ and some other open set O'.

One repeatedly applies the above to points $x' \in O'$, etc., showing that the entire open set is a union of intervals as desired. Finally, we remark that the set of these open intervals is either finite or at worst a countable infinite set. For (by axiom of choice) there is a rational number in each open interval, so the set of all these open intervals is in 1-1 correspondence with a subset of the integers (since the rational numbers are in 1-1 correspondence with the integers). But one easily sees that a subset of the integers is either finite, or in 1-1 correspondence with the integers (because one can index the subset by the places in which the elements occur, under the usual order of the integers).

Problems

1. If d is a metric on a set S, show that

$$d_1(x, y) = \frac{d(x, y)}{1 + d(x, y)}$$

and $d_2(x, y) = $ minimum $(d(x, y), 1)$ are both metrics on S.

2. Two metrics d_1 and d_2 on S are equivalent if given any x, whenever we have $\epsilon_2 > 0$, there is $\epsilon_1 > 0$ so that

$$\{y \mid y \in S, d_1(x, y) < \epsilon_1\} \subseteq \{y \mid y \in S, d_2(x, y) < \epsilon_2\}$$

and whenever $e_1^1 > 0$, there is $e_2^1 > 0$ so that

$$\{y \mid y \in S, d_2(x, y) < \epsilon_2^1\} \subseteq \{y \mid y \in S, d_1(x, y) < \epsilon_1^1\}$$

A) Show that any metric d is equivalent to the metric $\alpha \cdot d$, where $\alpha \in \mathbb{R}, \alpha > 0$.

B) Show that d, d_1, and d_2 in Problem 1 are all equivalent.
3. Show that the following set in the plane is open:
$$\{(x, y) \mid (x, y) \in E^2, x > 0, y > 0, xy < 1\}.$$
4. Show that the following are *not* metrics on E^2.
   A) $d_1((x_1, y_1), (x_2, y_2)) = |x_1 - x_2|$
   B) $d_2((x_1, y_1), (x_2, y_2)) = |x_1 y_1 - x_2 y_2|$
   C) $d_3((x_1, y_1), (x_2, y_2)) = x_1 + x_2 - y_1 - y_2$.
5. Show that the following subsets of the plane are *not* open
   A) The x axis
   B) $\{(x, y) \mid x^2 + y^2 \leq 1\}$
   C) $\{(x, y) \mid xy \geq 0\}$.
6. Let S be a metric space with metric d, $x_0 \in S$ a fixed point or element of S.

   Define a function from S to the real numbers by
   $$f(x) = d(x_0, x).$$
   Prove that f is a continuous function.
7. Let f, g be two continuous functions from a metric space S to the real numbers.
   Define
   $$h(x) = \max\ (f(x), g(x)),$$
   that is, $h(x)$ is $f(x)$ when $f(x) \geq g(x)$, otherwise $g(x)$.
   Prove that $h(x)$ is continuous.
8. Prove, showing only that the inverse image of an open set is an open set, that the following are continuous.
   A) $f: \mathbb{R} \to \mathbb{R}$ is defined by $f(x) = x^2 + 3$.
   B) $f: E^2 \to \mathbb{R}$ by $f(x, y) = x$.
   C) $f: E^2 \to \mathbb{R}$ by $f(x, y) = x^2 + y^2$.
9. What relationship is there between the sets $f^{-1}(V_1 \cap V_2)$ and $f^{-1}(V_1) \cap f^{-1}(V_2)$, where $f: S_1 \to S_2$ is a function, and V_1 and V_2 are subsets of S_2?
10. Using Proposition 2.9, show that if O is any open subset of $\mathbb{R}$, except ϕ or $\mathbb{R}$ itself, there is a point $z \in \mathbb{R} - O$, which has the property that for any $\epsilon > 0$, there is a point of O nearer to z than ϵ.

CHAPTER 3

General Topological Spaces. Bases. Continuous Functions. Product Spaces

We now come to the general theory of topological spaces, which is the traditional material of point-set or general topology. The need for a study of general spaces comes from the fact that the class of metric spaces is rather restrictive with respect to certain operations. For example, one could show that a product of a family of metric spaces need not be a metric space in the sense that there is no distance function which gives the expected open sets. But what is more important is the fact that many genuinely important properties do not depend on a specific metric, so that even for metric spaces, the restriction to a specific metric obscures the study of new and interesting concepts. In addition, there is a general need for determining when two spaces should be viewed as equivalent from the point of view of topology; this notion, as adopted by all modern mathematicians, is rather more general than to demand that all distances be the same (this latter being the more restrictive notion of "isometric"). Intuitively, a topological space is a simple idea; it consists of a set and a specified family of subsets, called the open sets, which satisfy the same sort of conditions as Proposition 2.8. We shall study the basic definitions first, and then look into bases and subbases, which are devices for specifying the structure of a topological space (called a topology, for short) in a minimum of terms. Then we shall study notions of subspace and the basic concept of product space. All of the material in this chapter is very basic and it has crept, by now, into almost all parts of mathematics.

Definition 3.1

A *topological space* X is a set, with a specified family of subsets θ (that is a collection of subsets $0_\alpha \subseteq X$, perhaps very large in number) so that
   TA) ϕ and X belong to the family θ, or in short $\phi \in \theta$ and $X \in \theta$.

TB) If we have any collection of sets $\{O_\beta\}$, $\beta \in B$, where B is the set of all indices β, and where each O_β belongs to θ (the family of open sets), then

$$\bigcup_{\beta \in B} O_\beta$$

also belongs to θ.

TC) If $O_1, \cdots, O_n$ are any finite number of open sets belonging to θ, then

$$O_1 \cap O_2 \cap \cdots \cap O_n \text{ belongs to } \theta.$$

The specified family of open sets is called the *topology* on X. In short, a topology is a family of open sets, containing ϕ and X, which is closed under the operation of union and finite intersection.

Remarks. 1) ϕ is always a topological space with no open sets. (This is as much a matter of notational convenience, as a genuine mathematical assertion.)

2) Any metric space is a topological space, with the open sets defined in Definition 2.10. Proposition 2.8 shows that our three conditions are satisfied.

3) Any non-empty set X can be made into a topological space in two essentially trivial ways.

a) The only open sets are ϕ and X. One trivially checks TA), TB), and TC). This topology (or family of open sets) is called *indiscrete*.

b) The open sets are all subsets of X. Or in the notation of Chapter 1, $\theta = P(X)$. This one is called *discrete*.

4) In general, a given set admits many different topologies, that is different families of open sets which all satisfy Definition 3.1.

5) Do not make the mistake of thinking that any (infinite) intersection of open sets will be open. Consider the open intervals (open sets) $\langle -1/n, 1/n \rangle$ where n runs through the positive integers. Then the intersection is

$$\langle -1, 1 \rangle \cap \langle -\tfrac{1}{2}, \tfrac{1}{2} \rangle \cap \langle -\tfrac{1}{3}, \tfrac{1}{3} \rangle \cdots = 0$$

which is *not* open.

6) If $X = \{a, b\}$, i.e. X consists of two elements, then X has the following topologies:

$$\phi, \{a, b\} \text{ (the indiscrete)},$$
$$\phi, \{a, b\}, a, b \text{ (the discrete)},$$
$$\phi, \{a, b\}, a, \text{ and}$$
$$\phi, \{a, b\}, b.$$

Now, it is clear that the crucial thing about a topological space is the open sets. But, there is a perfectly symmetrical notion of closed sets, as

Definition 3.2

A set $A \subseteq X$, X a topological space, is *closed*, if $X - A = \{x \mid x \in X, x \notin A\}$ is open.

Remarks. 1) Since $X = X - \phi$ and $\phi = X - X$, we see that ϕ and X are both open and closed.

2) The closed sets satisfy the conditions that any finite union of closed sets is closed, and any intersection of closed sets (even infinitely many) is closed.

Definition 3.3

If $A \subseteq X$, X a topological space, then the *interior* of A, written as A^i is defined as follows:

$$A^i = \bigcup_\beta O_\beta$$

where O_β runs through the collection of open sets O_β which are subsets of A.

Definition 3.4

As above, the *exterior* of A is the union of all open sets which do *not* meet A, written A^e, and defined formally as

$$A^e = \bigcup_\gamma O_\gamma$$

where O_γ runs through all open sets such that $O_\gamma \cap A = \phi$.

Definition 3.5

The *frontier* of A, or *boundary* of A, written A^b, consists of all points $x \in X$ with the property that if O is any open set with $x \in O$, then

$$O \cap A \neq \phi$$
$$O \cap (X - A) \neq \phi,$$

that is, O meets A and its complement, namely $X - A$.

Remarks. Clearly A^i, A^e, and A^b are disjoint, and their union is X. Also, A^i and A^e are the unions of open sets, and hence are open. We give various examples in R.

Let $A = \{x \mid x \in \mathbb{R}, a \leq x < b\}$. Then
$$A^i = \langle a, b \rangle, \quad A^e = \{x \mid x \in \mathbb{R}, x < a \text{ or } b < x\}$$
$$A^b = \{a, b\}, \text{ the set consisting of } a \text{ and } b.$$

Let X be any topological space
$$\phi^i = \phi, \quad \phi^e = X, \quad \phi^b = \phi.$$
$$X^i = X, \quad X^e = \phi, \quad X^b = \phi.$$

If $B = \{x \mid x \in \mathbb{R}, x \geq 0\}$, then
$$B^i = \{x \mid x \in \mathbb{R}, x > 0\}, \quad B^e = \{x \mid x \in \mathbb{R}, x < 0\}, \quad B^b = \{0\}.$$

The verifications of these assertions are all easy, but the reader should check some of them to be sure of understanding the concepts. Further illustrations are in the exercises following this section.

These concepts are not unrelated, but, before exploring the relations, we should give one further definition and one basic proposition.

Definition 3.6

Let $A \subseteq X$, X a topological space.
Define the closure of A, written $\bar{A}$, by
$$\bar{A} = A^i \cup A^b.$$

Note that if $A \subseteq B$, $A^i \subseteq B^i$ and $\bar{A} \subseteq \bar{B}$ follow at once from the definitions.

Now, clearly $\bar{A} = X - A^e$, but a more useful interpretation is given in the following:

Proposition 3.1

Let $A \subseteq X$, X being a topological space.
Then
$$\bar{A} = \bigcap_\alpha C_\alpha$$
where C_α runs over all closed sets which contain A. In other words, $\bar{A}$ is the set of points which belong to all closed sets containing A, or it is the smallest closed set containing A.

Proof. We first show $\bar{A} \subseteq \bigcap_\alpha C_\alpha$. Let C_α be a closed set containing A. Clearly, $A \subseteq C_\alpha$, but then $A^i \subseteq C_\alpha$ as $A^i \subseteq A$. We also will show that $A^b \subseteq C_\alpha$. If $x \in A^b$, every open 0, with $x \in 0$ meets A and $X - A$. Now, if $x \in A^b$, but $x \notin C_\alpha$, then $x \in X - C_\alpha$ which is an open set not meeting C_α. As $A \subseteq C_\alpha$, then $X - C_\alpha$ does not meet A, contradicting the definition of A^b. Hence, whenever $x \in A^b$, then $x \in C_\alpha$, or $A^b \subseteq C_\alpha$.

We have proved that for any α whatsoever,
$$\bar{A} = A^i \cup A^b \subseteq C_\alpha.$$
Thus, $\bar{A}$ belongs to all C_α at once, or
$$\bar{A} = A^i \cup A^b \subseteq \bigcap_\alpha C_\alpha,$$
as asserted.

Next, note that $\bar{A}$ is closed (check) and $A \subseteq \bar{A}$. Thus $\bar{A}$ is one of the C_α's, say C_{α_0}. Then
$$\bigcap_\alpha C_\alpha = C_{\alpha_0} \cap \bigcap_{\alpha \neq \alpha_0} C_\alpha = \bar{A} \cap \bigcap_{\alpha \neq \alpha_0} C_\alpha \subseteq \bar{A}.$$

Thus, we must have $x \in \bar{A}$, whenever $x \in \bigcap_\alpha C_\alpha$; we have concluded that
$$\bigcap_\alpha C_\alpha = \bar{A}$$
as asserted. I recommend drawing a series of pictures, such as

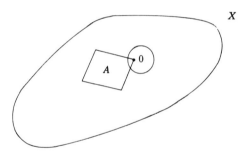

(here representing a frontier or boundary point) to check each step of the proof.

We now study the definitions which we have made.

Proposition 3.2

Let $A, B \subseteq X$, X a topological space
1) $(X - A)^i = A^e$
2) A is open if and only if $A = A^i$
3) A is closed if and only if $A = \bar{A}$
4) $\bar{A} \cup \bar{B} = \overline{A \cup B}$
5) $(A \cap B)^i = A^i \cap B^i$
6) $\overline{A \cap B} \subseteq \bar{A} \cap \bar{B}$
7) $A^i \cup B^i \subseteq (A \cup B)^i$.

Proofs. 1) is clear because both sides mean those points which are not in A, and which lie in an open set which does not meet A.

2) If A is open and $x \in A$, take $A = 0$, so that
$$x \in 0 \subseteq A,$$
proving that $x \in A^i$. As $A^i \subseteq A$, we conclude that $A = A^i$. Conversely, if $A = A^i$, then as A^i is open, A must be open.

3) Clearly, if A is closed, the intersection of all closed sets which contain A is an intersection of sets including the set A. Thus
$$\bar{A} = \bigcap_\alpha C_\alpha = A \cap (\bigcap_\beta C_\beta)$$
where α runs through all the closed set containing A, and β runs through all those except A itself. But clearly
$$A \cap (\bigcap_\beta C_\beta) \subseteq A,$$
showing that $\bar{A} \subseteq A$. As $A \subseteq \bar{A}$ always, we are done. Conversely, since $\bar{A}$ is closed, if $A = \bar{A}$, then A is trivially closed.

4) $A \cup B \subseteq \bar{A} \cup \bar{B}$, which is a closed set; since $\overline{A \cup B}$ is the smallest closed set containing $A \cup B$, clearly
$$\overline{A \cup B} \subseteq \bar{A} \cup \bar{B}.$$
Conversely, $A \subseteq A \cup B$ so clearly $\bar{A} \subseteq \overline{A \cup B}$. A similar argument for B shows $\bar{B} \subseteq \overline{A \cup B}$. But then we have
$$\bar{A} \cup \bar{B} \subseteq \overline{A \cup B}$$
as desired.

5) The reader should check the easy fact that
$$A^i \cap B^i \subseteq (A \cap B)^i;$$
on the other hand, let $x \in (A \cap B)^i$. Then there is an open set 0 with
$$x \in 0 \subseteq A \cap B$$
as $A \cap B \subseteq A$ and $A \cap B \subseteq B$, we also have $x \in 0 \subseteq A$, which means $x \in A^i$ and $x \in 0 \subseteq B$, which means $x \in B^i$. Thus $x \in A^i \cap B^i$. We have shown that
$$(A \cap B)^i \subseteq A^i \cap B^i.$$

6) If a closed set C_1, contains A and a closed set C_2 contains B, then the closed set $C_1 \cap C_2$ contains $A \cap B$. From this, one easily checks that the intersection of the closed sets which contain A and the closed sets which contain B must necessarily contain the intersection of the closed sets which contain $A \cap B$. The reader should write down the details here.

7) Let $x \in A^i \cup B^i$, and say $x \in A^i$. Then there is an open 0 so that

$$x \in 0 \subseteq A \subseteq A \cup B$$

showing that $x \in (A \cup B)^i$.
This completes the proof. Although these basic facts are in themselves uninteresting, they form an important cornerstone of our subject. For this reason, the student should satisfy himself/herself thoroughly with their correctness, before proceeding.

It would certainly be just to accuse me of cramming too much into this initial section, if I were to add one more definition. Yet there is another definition which is especially appropriate here, so please bear with me. When we are done with that, we shall have mastered all the basic concepts about open and closed sets in a topological space.

Definition 3.7

Let $A \subseteq X$ be a topological space. Let $x_0 \in X$. x_0 is called an *accumulation point* for A, if whenever 0 is an open set, $x_0 \in 0$, then 0 contains at least one point which lies in A and is not the point x_0 itself.

An accumulation point of A is a "pile-up" of points in A, excluding the point itself if it happens to be in A. If it doesn't happen to lie in A, it's simply a "pile-up of points in A." The set of all accumulation points of A is written A' and sometimes referred to as the *derived set* of A.

Proposition 3.3

Let $A \subseteq X$, X being a topological space.
Then

$$\bar{A} = A \cup A'.$$

Proof. Note $\bar{A} = A^i \cup A^b$, $A^i \subseteq A$. We show now that $A^b \subseteq A \cup A'$. Let $x \in A^b$, $x \notin A$. Then for every open set 0 with $x \in 0$,

$$0 \cap A \neq \phi \quad \text{and} \quad 0 \cap (X - A) \neq \phi.$$

Select $x_1 \in 0 \cap A \subseteq 0$. As $x \notin A$, $x \neq x_1$. This proves that $x \in A'$. Of course, if $x \in A^b$ and $x \in A$, then $x \in A \cup A'$ trivially. Hence

$$\bar{A} \subseteq A \cup A'.$$

On the other side of things, $A \subseteq \bar{A}$, so we must show $A' \subseteq \bar{A}$. Let $x \in A'$. If $x \notin \bar{A}$, there is a closed set C with $x \notin C$, $A \subseteq C$. Then $x \in X - C$ which is open and does not meet A, contradicting the assumption that

$x \in A'$. Thus $A' \subseteq \bar{A}$, and putting these remarks together

$$A \cup A' \subseteq \bar{A},$$

completing the proof.

Corollary

$$A' \subseteq \bar{A}.$$

Warning. Many other assertions of the general type in Propositions 3.1, 3.2, and 3.3 have a superficial appearance of being true, or may be true in some cases but false in others. For example.

1) $A^i \cup B^i = (A \cup B)^i$ *is false*. Let

$$A = \{x \mid x \in \mathbb{R}, 0 \leq x \leq 1\}$$
$$B = \{x \mid x \in \mathbb{R}, 1 \leq x \leq 2\}$$

$A^i = \langle 0, 1 \rangle$, $B^i = \langle 1, 2 \rangle$, $A \cup B = [0, 2]$, $(A \cup B)^i = \langle 0, 2 \rangle$.

Thus, $A^i \cup B^i$ does not contain 1, while $(A \cup B)^i$ does.

2) $A' = A^b$ is false. In fact, both $A' \subseteq A^b$ and $A^b \subseteq A'$ are false in general. Let $A = X = \mathbb{R}$. Then $A' = \mathbb{R}$, but $A^b = \phi$, because no point of $\mathbb{R}$ meets $A = \mathbb{R}$ and its complement. Hence, A' is not contained in A^b.

Let $A = \{0\} \subseteq \mathbb{R}$. Then $A^b = A$. But $A' = \phi$ because no open sets containing $\{0\}$ meets another point of A, and every point other than zero is contained in a (perhaps small) open set which does not contain zero. Thus, A^b is not necessarily contained in A'.

Care is needed here, and I urge the student, who is not already familiar with these topics, to work virtually all of the following problems. They are basically straightforward and will ensure an understanding.

Problems

1. Find the number of distinct topologies on $\{a, b, c\}$, the set with three elements.
2. Let $X = \mathbb{R}$, $A = \langle 0, 1 \rangle \cup \{2\} \cup [3, 4]$. Find explicitly, A^i, A^b, A^e, A', and $\bar{A}$.
3. Let $X = \mathbb{R}$, $A = \langle 0, 1 \rangle = \{x \mid x \in \mathbb{R}, 0 < x < 1\}$, $B = \langle -1, 0 \rangle$. Show that

$$\overline{A \cap B} = \phi$$

and

$$\bar{A} \cap \bar{B} = \{0\},$$

the set consisting of the single element 0. This shows that Proposition 3.2, Part 6, cannot be improved.

4. Prove that if $A \subseteq B \subseteq X$, X a topological space, then
   A) $A^e \supseteq B^e$
   B) $A' \subseteq B'$.
5. Show by example (subsets of $\mathbb{R}$ will suffice) that we may have
$$A \subseteq B$$
while
$$A^b \subseteq B^b \text{ is false.}$$
6. Prove that, with the above notations, A is closed, if and only if $A^b \subseteq A$.
7. Prove that A is closed, if and only if $A' \subseteq A$ (use Proposition 3.3).
8. Let A = rational numbers, X = real numbers. Show that
$$A^e = A^i = \phi.$$
9. Let $X = E^2$, the plane A = the x-axis. Find A^i, A^e, A^b and A'. Show also that A is closed.
10. Let X be a topological space. A subset $N \subseteq X$, with $x \in N$ is called a *neighborhood* of x, if there is an open 0 with
$$x \in 0 \subseteq N.$$
These neighborhoods have interesting properties, and may even be used to define topologies.

Show here the following easy statements:
   a) If N_1 and N_2 are neighborhoods of x, then so is $N_1 \cap N_2$.
   b) If N_1 is a neighborhood of x, and M is any subset containing N_1, then M is a neighborhood of x.
   c) If $x \in 0 \subseteq N$, defined as a neighborhood of x, and if y is any point in 0, then N is a neighborhood of y.
11. Let X be the subspace of E^2 consisting of $\{(x, y) \mid y \geq 0\}$ and the single point $(0, -1)$. Show that there is a ball $B_r((x_0, y_0))$ whose closure in X is strictly smaller than
$$\{(x, y) \mid d((x_0, y_0), (x, y)) \leq r\}.$$

Assuming that the previous material is well under your belt, we now will discuss bases and subbases, which are very useful devices for specifying the open sets in a topological space, without having to go to all the trouble of enumerating them all. In later sections, we shall need to look at topologies which have the fewest open sets, consistent with certain conditions, and it will be particularly useful to describe these explicitly in the language of bases.

Definition 3.8

Let X be a set, $\mathcal{B}$ a family of subsets. Suppose $\mathcal{B}$ satisfies the following two properties.

B1) Let $x \in X$. Then there is $B \in \mathcal{B}$ so that $x \in B$. In other words for every $x \in X$, there is at least one set in the family of subsets $\mathcal{B}$ which contains x.

B2) Let $x \in B_1$, $x \in B_2$, with $B_1, B_2 \in \mathcal{B}$. Then there is a $B_3 \in \mathcal{B}$ so that

$$x \in B_3 \subseteq B_1 \cap B_2.$$

This says, very simply, that if a point lies in the intersection of two sets in the family, it lies in a set in the family which is a subset of the intersection.

Such a family of sets, $\mathcal{B}$, is called a *basis*.

In fact, a *basis* gives rise to a topology and is frequently called a basis for a topology, as we now prove.

Proposition 3.4

If X is a set and $\mathcal{B}$ is a basis on X, then the family of sets made up of unions of sets in $\mathcal{B}$, that is the family of sets of the form

$$\bigcup_\alpha B_\alpha, \quad \text{with} \quad B_\alpha \in \mathcal{B},$$

for some (or possibly all) the α's, is a topology. $\mathcal{B}$ is called the basis for the topology.

Proof. The union of no sets is the empty set. As each x belongs to some B, the union of all the B's is X, so that axiom TA) for a topology is satisfied.

The union of a bunch of unions of sets, all in $\mathcal{B}$, is of course just a union (perhaps gigantic) of sets in $\mathcal{B}$. Thus, the second condition, TB) for a topology, is satisfied.

To prove condition TC), let the following be sets of the given form (any finite number of such sets).

$$\bigcup_\alpha B_\alpha, \quad \bigcup_\beta B_\beta, \quad \cdots, \quad \bigcup_\tau B_\tau$$

We must prove that

$$\bigcup_\alpha B_\alpha \cap \bigcup_\beta B_\beta \cap \cdots \cap \bigcup_\tau B_\tau$$

is actually a union of sets in $\mathcal{B}$. Take any x in this finite intersection. Such an x must of course lie in each component of the intersection, so that there are sets B_{α_1}, from the first union, B_{β_1} from the second union, etc., with

$$x \in B_{\alpha_1} \cap B_{\beta_1} \cap \cdots \cap B_{\tau_1}.$$

Here we mean that α_1 is some fixed α, β_1 some fixed β, and so on.

Applying axiom B2) for a basis, repeatedly, we quickly see that there must be $B \in \mathcal{B}$, with

$$x \in B \subseteq B_{\alpha_1} \cap \cdots \cap B_{\tau_1}.$$

Since $B_{\alpha_1} \subseteq \bigcup_\alpha B_\alpha$; $B_{\beta_1} \subseteq \bigcup_\beta B_\beta$, etc., clearly

$$x \in B \subseteq (\bigcup_\alpha B_\alpha) \cap (\bigcup_\beta B_\beta) \cap \cdots \cap (\bigcup_\tau B_\tau)$$

and we call, for convenience, the intersection of these unions C. Summarizing, we have proved that for every x belonging to

$$C = \bigcup_\alpha B_\alpha \cap \cdots \cap \bigcup_\tau B_\tau$$

there is some B in $\mathscr{B}$ which lies in this intersection and also contains x.

Take the union of all such sets $\mathscr{B}$. This union contains every x in C, because there is a set B for every x. Thus, the union of all such sets contains C. But the union of all such sets is also contained in C, because each such set B is contained in C.

The result is then that C is a union of sets from B, as desired. We have thus checked out the three conditions for a topology.

Some remarks and examples will surely be helpful here.

Remarks. 1) Let $\mathscr{R}$ be the real numbers, $\mathscr{B}$ the set of all open intervals. Since the intersection of two open intervals is either empty or an open interval, we see at once that $\mathscr{B}$ is a basis. Since any open set is the union of open intervals (in fact, we have proved the stronger assertion Proposition 2.9, though we don't need it here), the topology in question is the usual topology on the metric space $\mathscr{R}$.

2) We have used the following trivial assertion which is commonplace but deserves attention:

If $\mathscr{B}$ is a family of sets $\{B_\alpha\}$ in a set X, satisfying the properties.

a) Every $x \in X$ belongs to some B_α.

b) Every B_α is a subset of X.

Then $\bigcup_\alpha B_\alpha = X$. The proof is clear, because whenever b) holds, the union is a subset of X, while whenever a) holds, X must be a subset of the union.

3) If θ is a topology on a topological space X, then θ is always a basis. (Check this.) Topologies are often bigger. Bases usually offer the advantages of a more simple way of describing a topology.

We pass to subbases, which are, in an intuitive sense, crude families of subsets which generate bases (which then, in turn, generate topologies, as we have just seen).

Definition 3.9

A family of subsets $\mathcal{S}$ of a set X is called a *subbasis*, if
SB1) Every $x \in X$ belongs to at least one set of the family $\mathcal{S}$.

Proposition 3.5

If S is a subbasis on the set X, then the sets

$$B_1 \cap \cdots \cap B_n,$$

which are finite intersections (of any finite length) of sets $B_i \in S$, form a basis on X.

Proof. Let $B_1 \cap \cdots \cap B_n$ and $C_1 \cap \cdots \cap C_m$ be two finite intersections of sets in S. Suppose

$$x \in B_1 \cap \cdots \cap B_n \cap C_1 \cap \cdots \cap C_m = D$$

calling the entire intersection D. Then we have, trivially

$$x \in D \subseteq (B_1 \cap \cdots \cap B_n) \cap (C_1 \cap \cdots \cap C_m)$$

where the inclusion relation is actually an equality here. Since D is a set of the type in question, axiom B2) for a basis is satisfied. Axiom B1) is identical to SB1), so the proof is complete.

As an example, let S consist of the following subsets of R.

$$\langle a, \infty \rangle = \{x \mid x \in R, x > a\}$$
$$\langle -\infty, b \rangle = \{x \mid x \in R, x < b\}.$$

One easily checks that this is a subbasis, and the basis which it gives rise to is the basis of open intervals discussed above.

We now wish to compare two topologies on a given set.

Definition 3.10

Let θ_1 and θ_2 be two topologies on a set X (that is both θ_1 and θ_2 are families of subsets which satisfy the TA), TB), and TC) of Definition 3.1).

Say θ_1 is *finer* than θ_2 if every open set of θ_2 is an open set of θ_1, although θ_1 may have other open sets. The finer the topology, the more the open sets.

Remarks and Examples. 1) The discrete topology (every subset open) is finer than any other topology on a given set.

2) The indiscrete topology is coarser than any other topology.

3) The sets $[a, b\rangle = \{x \mid x \in R, a \leq x < b\}$ form a basis on R. (Check.) The corresponding topology is the half-open interval topology. It is coarser than the discrete topology, because single points are open in the discrete topology but no single point is a union of half-open intervals.

On the other hand, it is finer than the usual topology defined by the metric, because every open interval $\langle a, b \rangle$ is the union of all the half-open intervals

$$\bigcup_{a < \gamma < b} [\gamma, b\rangle$$

(check!), but no $[c, d\rangle$ is a union of open intervals, for if c were in one open interval, we would be forced to include some points less than c. Or, in other words, if a point c lies in an open interval, that interval contains points to the left and to the right of c.

4) The *finite complement topology* on an infinite set X. For this topology, the non-empty open sets are of the form

$$0 = X - F$$

where F is finite.

This is easily checked to be coarser than the discrete but finer than the indiscrete topology.

5) If X is a set with a topology θ and a basis $\mathcal{B}$, then in order that the topology θ be coarser than the topology given by the basis $\mathcal{B}$, it suffices that for every $0 \in \theta$, $x \in 0$, there is $B \in \mathcal{B}$ with

$$x \in B \subseteq 0.$$

6) Naturally if θ_1 is finer than θ_2, and θ_2 finer than θ_1, we say that θ_1 and θ_2 are *equivalent topologies*.

Problems

1. Show that in a metric space X, the collection of open balls

$$B_\epsilon(x) = \{y \mid y \in X, d(x, y) < \epsilon\}$$

is a basis for the topology defined earlier (Definition 2.10).

2. Let X be an infinite set. Let $\mathcal{S}$ be the collection of subsets which are the complements of single points, i.e. a member of $\mathcal{S}$ looks like

$$X - \{a\}.$$

Show that $\mathcal{S}$ is a subbasis for the topology of finite complements (Example 4 above).

3. Show that the topology of finite complements is the coarsest topology on any set X for which single points are closed sets.

4. Show that the following sets form a basis for the topology on the plane E^2:

$$\left\{(x,y) \,\middle|\, \begin{array}{l} a<x<b \\ c<y<d \end{array}\right\}$$

(*Hint:* Show that it is a basis first. Then show that a set is open for this topology if and only if it is open in the sense of Definition 2.10. This comes down to showing that if (x, y) lies in a rectangle (without its edges) there is an open disc about (x, y) lying entirely in the rectangle, and if (x, y) lies in some disc, then there is a rectangle containing it, lying within the disc.)

Topology

Finally, we come to continuous functions, subspaces, and products for general topological spaces. Our general definition is motivated by the basic Theorem 2.1 of the previous chapter.

Definition 3.11

Let X_1 and X_2 be topological spaces and $f: X_1 \to X_2$ a map (that is f assigns to each $x \in X_1$, a single $f(x)$ in X_2).

f is said to be *continuous*, if whenever $0 \subseteq X_2$ is an open set, $f^{-1}(0) \subseteq X_1$ is an open set. (Recall that $f^{-1}(0)$ is the set all $x \in X_1$, with $f(x) \in 0$.)

Remarks. 1) Theorem 2.1 shows that this definition agrees with the more traditional definition given in *Definition 2.9(B)* in the case of metric spaces.

2) If X is any topological space, $1: X \to X$ defined by $1(x) = x$ is trivially checked to be continuous. 1 is usually referred to as the identity function.

3) If $f: X_1 \to X_2$ and $g: X_2 \to X_3$ are continuous functions, between topological spaces, then $g \cdot f: X_1 \to X_3$ (defined as $g \cdot f(x) = g(f(x))$) is continuous. To prove this, note that $(g \cdot f)^{-1}(0) = f^{-1}(g^{-1}(0))$. (Check!) Thus, if 0 is open and g is continuous, $g^{-1}(0)$ is open. But if f is continuous, then $f^{-1}(g^{-1}(0))$ is also open, so $(g \cdot f)^{-1}(0)$ is open as desired.

4) Let X_1 and X_2 be the same set, but with different topologies, and suppose the topology on X_2 is coarser than that on X_1, i.e. fewer open sets Then

$$1 : X_1 \to X_2, \quad 1(x) = x$$

is continuous, because $1^{-1}(0) = 0$, and an open set in X_2 is assumed to be open in X_1.

On the other hand, if the topology on X_2 is strictly finer than that on X_1, there would be an open set whose inverse image is not open, and 1 would be discontinuous (or not continuous).

If $A \subseteq X$ is a subspace of a topological space, is it itself a topological space in a reasonable way? Proposition 2.6 makes us look into this question, and the following definition affords an affirmative answer.

Definition 3.12

Let X be a topological space, $A \subseteq X$ a subset. Define a family θ_A of subsets of A to be all subsets

$$A \cap 0$$

where 0 is an open set in X.

Clearly, θ_A contains the empty set and A. But, the obvious relations (check)

$$A \cap (\bigcup_\alpha 0_\alpha) = \bigcup_\alpha (A \cap 0_\alpha)$$

and $A \cap (0_1 \cap \cdots \cap 0_n) = (A \cap 0_1) \cap (A \cap 0_2) \cdots \cap (A \cap 0_n)$ show that unions and finite intersections of sets of this form also have this form.

Thus, the family of subsets of A, θ_A, is a topology on the set A. It is called the *relative* or *subspace* topology for the subset A of the topological space X. A is often called a *subspace* of X.

This topology turns out to be an optimal topology on a subset of a space, because, in a sense that we will make clear below, it is just the right minimal number of open sets to reflect the fact that A lives in X.

Proposition 3.6

Let X be a topological space, A a subset, given the relative topology. Let
$$i: A \to X$$
be the inclusion map, defined by $i(x) = x$ for any $x \in A$. Then,

1) i is continuous.

2) If A is given any other topology or family of open sets, which is strictly coarser (less open sets) than the relative topology, then i is discontinuous.

Proof. Let $0 \subseteq X$ be open. One checks immediately that
$$i^{-1}(0) = A \cap 0.$$
Hence, because $A \cap 0$ is defined to be an open set in A, i must be continuous.

To prove the second assertion, note that if the topology on A is coarser than the relative topology, then there is some set
$$A \cap 0_1$$
which is not open in that topology, even though $0_1 \subseteq X$ is open. But then
$$i^{-1}(0_1) = A \cap 0_1$$
fails to be open and i is discontinuous.

Remark. This proposition actually shows that the relative or subspace topology is the coarsest topology on A so that the inclusion map is continuous. In other words, it is the most economical way to make A a topological space, consistent with an obvious map being continuous. This general procedure of defining a topology to be the most economical, consistent with certain conditions, is a basic device in topology. The following definition of the topology on a cartesian product is motivated by these conditions.

Recall now that if $\{X_\alpha\}$ is a family of sets, where the collection of all the α's is written A, the Cartesian product,
$$\underset{\alpha \in A}{\times} X_\alpha$$

is defined as the set of all functions

$$f: A \to \bigcup_{\alpha \in A} X_\alpha$$

which has the property that $f(\beta) \in X_\beta$ whenever $\beta \in A$. One may think intuitively of sequences (in a general sense) of elements indexed according to the α's, with entries in the respective X_α's.

Associated with the Cartesian product, we have maps

$$\pi_\beta: \underset{\alpha \in A}{\times} X_\alpha \to X_\beta, \quad \text{for any fixed} \quad \beta \in A,$$

defined as $\pi_\beta(f) = f(\beta)$. These are frequently called the *projections* onto the factor X_β. Our goal is to show that when the X_α's are also topological spaces, we can make the Cartesian product a topological space in a coarsest possible way, consistent with the projection maps being continuous.

Definition 3.13

Let $\{X_\alpha\}$ be a family of topological spaces, $\alpha \in A$.
Define a subset

$$B(\beta, 0_\beta) = \{f \mid f \in \underset{\alpha \in A}{\times} X_\beta, f(\beta) \in 0_\beta\}$$

where $0_\beta \subseteq X_\beta$ is an open set. The sets $B(\beta, 0_\beta)$ for all β and 0_β clearly form a subbasis. The associated basis gives rise to a topology, which is called the *product topology* on the Cartesian product:

$$\underset{\alpha \in A}{\times} X_\alpha.$$

To get some insight into this definition, we need a clearer description of just what is an open set in the Cartesian product.

Proposition 3.7

An open set in $\times_{\alpha \in A} X_\alpha$ is a union of sets of the form

$$\{f \mid f \in \underset{\alpha}{\times} X_\alpha, f(\alpha_1) \in 0_{\alpha_1}, \cdots, f(\alpha_m) \in 0_{\alpha_m}\}$$

where $0_{\alpha_1} \subseteq X_{\alpha_1}, \cdots, 0_{\alpha_m} \subseteq X_{\alpha_m}$ are a finite collection of open sets in the respective spaces.

Proof. The proof is immediate, because a set of this form is a finite intersection of sets $B(\beta, 0_\beta)$. Thus, the sets which we are considering in this proposition are precisely the sets in the basis associated with our subbasis (see Proposition 3.5). The assertion is now clear (see Proposition 3.4); the topology comes from taking unions of sets in the basis.

Our earlier claims are now substantiated by the following:

Proposition 3.8

The product topology is the coarsest topology on $\times_{\alpha \in A} X_\alpha$ so that all π_β are continuous.

Proof. $\pi_\beta^{-1}(0_\beta) = B(\beta, 0_\beta)$ because, if $f \in \pi_\beta^{-1}(0_\beta)$, $f(\beta) \in 0_\beta$ and so $\pi_\beta(f) = f(\beta) \in 0_\beta$. If $f \in B(\beta, 0_\beta)$, it's clear that $f \in \pi_\beta^{-1}(0_\beta)$.

Thus, all π_β are continuous, because the inverse images of open sets are open sets (in fact even sets of the subbasis).

If we had a coarser topology, some union of open sets of the form

$$\{f \mid f \in \times_{\alpha \in A} X_\alpha, f(\alpha_1) \subseteq 0_{\alpha_1}, \cdots, f(\alpha_m) \in 0_{\alpha_m}\}$$

would fail to be open. Hence, one such set would fail to be open. But notice,

$$\{f \mid f \in \times_{\alpha \in A} X_\alpha, f(\alpha_1) \in 0_{\alpha_1}, \cdots, f(\alpha_m) \in 0_{\alpha_m}\} = \bigcap_{i=1}^{m} B(\alpha_i, 0_{\alpha_i})$$

where we mean the intersection of the m sets as i ranges from 1 to m. If all these were open, then our original set would be open, all in this coarser topology.

Hence, some $B(\alpha_j, 0_{\alpha_j})$ is *not* open in this topology. As

$$\pi_{\alpha_j}^{-1}(0_{\alpha_j}) = B(\alpha_j, 0_{\alpha_j})$$

(see the first part of the proof), π_{α_i} then would fail to be continuous.

Remarks. The upshot of all this is that the product topology is the coarsest topology which makes all projections continuous.

As our final topic for the chapter, we discuss the notion of two topological spaces being equivalent. This definition is modeled after similar notions in set theory, group theory, etc., and forms a special case of the more general notion of equivalent objects in a category.

Definition 3.14

Let X and Y be topological spaces. We say that X and Y are *equivalent* or *homeomorphic*, if there are two continuous maps

$$f: X \to Y$$

and

$$g: Y \to X$$

so that $g \cdot f = 1_X$ and $f \cdot g = 1_Y$ where the notation 1_X means the identity map from X to itself and ditto 1_Y.

In other words, the maps f and g are inverses of one another (exactly as in the section on 1-1 correspondences of Chapter 1, but requiring that the map be continuous). f or g is then called a *homeomorphism*.

Examples. 1) Any two spaces, both having the discrete topology whose elements are in 1-1 correspondence, are homeomorphic. In this case the maps which are inverses to one another, which define the 1-1 correspondence, are automatically continuous.

2) Let
$$X = \langle 0, 1 \rangle \subseteq \mathbb{R}$$
and
$$Y = \langle 1, \infty \rangle \subseteq \mathbb{R},$$
(recall that $\langle 1, \infty \rangle = \{x \mid x \in \mathbb{R}, x > 1\}$) both having the relative topology of subsets of $\mathbb{R}$, or equivalently, the topology defined by the metric $d(u, v) = |u - v|$. Then X and Y are homeomorphic.

Define
$$f: X \to Y \text{ by } f(x) = \frac{1}{x}$$
and
$$g: Y \to X \text{ by } g(y) = \frac{1}{y}$$

It is trivial to check that these maps are continuous, as 0 is *not* a member of X or Y. But
$$(g \cdot f)(x) = g(f(x)) = g\left(\frac{1}{x}\right) = \frac{1}{\frac{1}{x}} = x$$
proving $g \cdot f = 1_X$.

The same argument shows that $f \cdot g = 1_Y$, giving the desired result.

3) If $f: X \to Y$ is a homeomorphism, f (as well as the inverse g) must be a 1-1 correspondence. For suppose
$$f(x_1) = f(x_2).$$
Then
$$(g \cdot f)(x_1) = (g \cdot f)(x_2)$$
or
$$1_X(x_1) = 1_X(x_2)$$
or
$$x_1 = x_2$$
proving f is 1-1.

If f is not onto, say y is not an $f(x)$, then we cannot have
$$y = (f \cdot g)(y) = f(g(y));$$
thus f must be onto.

We conclude that any two spaces, whose elements are *not* in 1-1 correspondence, can't possibly be homeomorphic.

Further examples are in the problems.

In short, if two spaces are homeomorphic, they are virtually indistinguishable from the point of view of topology.

Problems

1. Let the real numbers have the topology with basis, the sets
   $$[a, b\rangle = \{x \mid x \in \mathbb{R}, a \leq x < b\}.$$
   Which of the following maps are continuous?
   a) $f(x) = x - 5$
   b) $f(x) = -x$
   c) $f(x) = 2x + 1$.
2. Let X and Y be topological spaces, $f: X \to Y$ a function (not assumed to be continuous). Let $\mathcal{G}$ be a subbasis for the topology on Y.
   Suppose that if $0 \in \mathcal{G}$, $f^{-1}(0)$ is open in X. Prove that f is continuous. (See Problem 9 at the end of Chapter 2).
3. Let $A \subseteq X$ be a closed set of a topological space X. Let $B \subseteq A$ be a subset of A.
   Prove that B is closed as a subset of A, if and only if B is closed as a subset of X.
4. Show that the previous Exercise 3 is false, if we omit the assumption that A is closed. (*Hint:* One easy example may be obtained by letting A be the union of two disjoint open intervals in the real line.)
5. Let X and Y be spaces and $f: X \to Y$ a function or map. Suppose that whenever $C \subseteq Y$ is closed, $f^{-1}(C) \subseteq X$ is closed. Prove that f is continuous.
6. Let X and Y be metric spaces. Then Proposition 2.7 and Definition 3.13 give two ostensibly different ways of describing a topology on $X \times Y$. (Recall that, whenever we have a metric space, Proposition 2.8 shows that we get a topology from Definition 2.10.) Prove that these topologies are the same, that is a subset of $X \times Y$ is open according to one definition, if and only if it is open according to the other. (*Hint:* This may be long, but it is not deep. If you think in terms of the plane, the open sets in the Proposition 2.7 have as a basis the interiors of circles, while in our new definition, they are the interiors of rectangles (at least as a basis). Draw a picture!)
7. Find a topological space X so that X and $X \times X$ are homeomorphic. (*Hint:* ϕ is an easy answer. Try for a more interesting one, at first using discrete topologies.)
8. Suppose X is the union of two disjoint open sets. (For example, X has two points and the discrete topology.) Suppose Y is homeomorphic to X.
   Show that Y is then the union of two disjoint, open sets.
9. Show that if $X - x$, $x \in X$, is the union of two disjoint, open sets, and if X is homeomorphic to Y, then there is a point y with $Y - y$ being the union of two disjoint, open sets.
10. Using Examples 7 and 8, show that the line and the plane are *not*

homeomorphic. (*Hint:* The line, with any point removed, is the union of two disjoint, open sets.)
11. Prove that the real line is homeomorphic to the open interval $\langle 0, 1 \rangle$, or in fact any open interval. (*Hint:* Build the homeomorphism out of such functions as $\tan(x)$, with suitable multiples.)

CHAPTER 4

The Special Notions of Compactness and Connectedness

Over the years, mathematicians have focused attention on a variety of special properties of topological spaces, which have occurred over and over again in a natural setting. For example, much interest lies in topologies which are rich in open sets, and one way this arises is for a space to have the property that different points live in non-overlapping open sets. Another property says, very crudely, that points do not string out infinitely far. Yet another property of importance is that the space is not made up of two essentially disjoint pieces.

All these properties, which we shall study here, have been shown to arise naturally in most branches of mathematics, including algebra and abstract analysis. They form a foundation stone in the knowledge of any serious student of mathematics.

Definition 4.1

A topological space X is called a *Hausdorff space*, if whenever $x, y \in X$, $x \neq y$, there are open sets 0_x and 0_y in X, with $x \in 0_x, y \in 0_y$,

$$0_x \cap 0_y = \phi.$$

In other words, distinct points lie in non-overlapping open sets. This notion was first recognized as fundamental by the German mathematician F. Hausdorff.

There are many easy examples. If X has the discrete topology, one checks trivially that X is Hausdorff. But much more depth is in the following:

Proposition 4.1

Every metric space is a Hausdorff space.

Proof. Let d be the metric on X. Suppose $x, y \in X$. Set, for these distinct x and y,

$$d(x, y) = \epsilon > 0.$$

Let $0_x = B_{\epsilon/3}(x) = \{z \mid z \in X, d(z, x) < \epsilon/3\}$ and $0_y = B_{\epsilon/3}(y)$.

I claim that these two open sets, which obviously contain x and y, respectively, are disjoint. For if $z \in 0_x \cap 0_y$, we have

$$d(x, z) < \epsilon/3, d(y, z) < \epsilon/3,$$

from which it follows that

$$\epsilon = d(x, y) \leq d(x, z) + d(z, y) < \tfrac{2}{3}\epsilon$$

or $1 < \tfrac{2}{3}$, which is absurd.

The proof may be represented, by the way of a sketch:

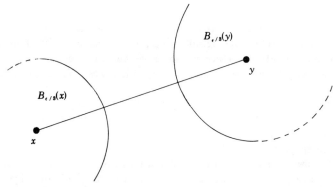

There are, however, plenty of examples of spaces, which are not metric (that is there is no distance function d, which gives use to the given topology), but are still Hausdorff. This will become clear in the following chapter, where we study conditions for a space to be a metric space.

For further examples, we have

Proposition 4.2

If X_α is a family of Hausdorff spaces, $\times_\alpha X_\alpha$ is a Hausdorff space.

Proof. Let $f, g \in \times_\alpha X_\alpha$. If f and g are different points of the product, there has to be some β with $f(\beta) \neq g(\beta)$. In X_β, which is Hausdorff by assumption, we find open sets 0_f and 0_g, so that $f(\beta) \in 0_f$, $g(\beta) \in 0_g$ and

$$0_f \cap 0_g = \phi.$$

But then, the open sets
$$B(\beta, 0_f) = \{h \mid h \in \times_\alpha X_\alpha, h(\beta) \in 0_f\}$$
and
$$B(\beta, 0_g) = \{h \mid h \in \times_\alpha X_\alpha, h(\beta) \in 0_g\}$$
are disjoint open sets in the Cartesian product (check!) and they contain f and g, respectively.

Remarks. 1) There are a wide variety of different notions of separation, of which the notion of a Hausdorff space is the most common. In many texts, a Hausdorff space is called a T_2-space. Hausdorff spaces are by far the most common in topology, but non-Hausdorff spaces do arise, in a natural way, in algebraic geometry, sheaf theory, etc.

2) In a Hausdorff space, a single point is closed. For consider $X - x_0$. For each $y \in X - x_0$, there are open $0_{x_0}, 0_y$ which contain x_0 and y and do not meet. Then $X - x_0$ is the union of all the open 0_y. (Check details here!)

Definition 4.2

A topological space X is called *compact*, if whenever $\{0_\alpha\}$ is a family of open sets in X with the property that
$$X = \bigcup_\alpha 0_\alpha,$$
there are a finite number of the 0_α's, let's say $0_{\alpha_1}, \cdots, 0_{\alpha_m}$, so that
$$X = 0_{\alpha_1} \cup 0_{\alpha_2} \cup \cdots \cup 0_{\alpha_m}.$$
In other words, if a family of open sets cover X, in the sense that their union is all of X, then a finite subfamily of this family also covers X.

Remarks and Examples. 1) A single point space is trivially compact.

2) The real line is *not* compact. For, consider the infinite family of open intervals
$$0_n = \langle n - \tfrac{3}{4}, n + \tfrac{3}{4}\rangle$$
for every integer n. Clearly, $\bigcup_n 0_n = \mathbb{R}$. But each integer $m \in \mathbb{R}$ is contained in exactly one such set, namely 0_m.

Thus, if we omit any 0_m whatsoever, the union of the rest fails to cover $\mathbb{R}$. In particular, no finite subfamily can cover $\mathbb{R}$.

3) A closed interval of real numbers, given the relative topology, is compact. This is the content of the famous Heine-Borel Theorem in analysis, and it will be proved here in the course of Theorem 4.1 below.

To gain some insight into compactness, we note the following:

Proposition 4.3

A necessary and sufficient condition that a topological space X be compact, is that when C_α is a family of closed sets in X, such that whenever $\alpha_1, \cdots, \alpha_n$ are a finite collection of α's, we have

$$C_{\alpha_1} \cap \cdots \cap C_{\alpha_n} \neq \phi,$$

then we have

$$\bigcap_\alpha C_\alpha \neq \phi.$$

In other words, if a family of closed sets has the property that any finite subcollection has at least one point in common, then all the sets have a point in common.

Proof. Let us assume X is compact. Put $0_\alpha = X - C_\alpha$, that is 0_α is the complement of C_α in X. Suppose, for a contradiction

$$\bigcap_\alpha C_\alpha = \phi.$$

I claim that the 0_α's cover X; for if $x \in X$ lies in no 0_α, it must lie in every C_α. But since X is compact, some finite subset of the 0_α's covers X, say

$$X = 0_{\beta_1} \cup \cdots \cup 0_{\beta_k}$$

where the $\beta_1, \cdots, \beta_k$ are some of the α's (a finite subcollection).

But then,

$$C_{\beta_1} \cap \cdots \cap C_{\beta_k} = \phi,$$

for if $y \in C_{\beta_1} \cap \cdots \cap C_{\beta_k}$, $y \in C_{\beta_i}$ for each i; or in other terms, $y \notin 0_{\beta_i}$ for each i.

Conversely, suppose our condition is satisfied, and $\{0_\alpha\}$ is a cover of X. We define $C_\alpha = X - 0_\alpha$, and note that

$$\bigcap_\alpha C_\alpha = \phi.$$

If our condition is assumed, however, this can only happen when some

$$C_{\alpha_1} \cap \cdots \cap C_{\alpha_m} = \phi.$$

But as above, this implies immediately that

$$0_{\alpha_1} \cup \cdots \cup 0_{\alpha_m} = X.$$

In short, the operation of complement takes unions into intersections, closed sets into open, and it takes the definition of compactness into the condition given in this proposition.

Examples. 1) In the real line $\mathbb{R} = E^1$, set
$$C_n = \{x \mid x \in \mathbb{R}, x \leq n\}$$
for all integers n (including negative).

Then any finite intersection of the sets is non-empty, because it will contain some small (very negative) numbers.

But no number is in all C_n, for every n. This, we see once again that $\mathbb{R}$ is not compact.

2) Let X be a topological space which is a finite set of points. Then, clearly X is compact, because one can never have an intersection of infinitely many closed subsets (there are only finitely many subsets of such a finite set). Thus our condition is trivially satisfied here.

We now head towards the important Theorem 4.1 which completely characterizes compact subsets in Euclidean spaces. In order to prove this, we call attention to two lemmas, which play an important role in many results to follow. These are easy to prove, but should be studied carefully.

Basic Lemma 4.1

a) Let X be a compact topological space. Let $A \subseteq X$ be closed. Then A, with the relative topology, is compact.

b) Let B be a subset of a Hausdorff space X, which is compact in the relative topology. Then B is closed in X.

Proof. a) Let $\{0_\alpha\}$ be a cover of A by open sets. By definition of the relative topology, we have
$$0_\alpha = A \cap U_\alpha$$
where U_α is an open set in X, for every α.

Consider the collection of sets consisting of all the 0_α's and $X - A$ in addition. This covers X, as any point is either in A, in which case it belongs to some 0_α, or not, in which case it belongs to $X - A$.

The collection of sets U_α, and $X - A$, in addition, then will surely cover X, because these sets are if anything bigger than the previous collection, which covers, i.e. $0_\alpha \subseteq U_\alpha$ for each α.

But as these are open in X and X is compact, there are a finite number, say
$$U_{\alpha_1}, \cdots, U_{\alpha_m}, X - A$$
which must cover X.

But then the $U_{\alpha_1}, \cdots, U_{\alpha_m}$ must cover A, as $X - A$ doesn't meet A. Then clearly, the sets
$$0_{\alpha_1} = A \cap U_{\alpha_1}, 0_{\alpha_2}, \cdots, 0_{\alpha_m}$$
must cover A, completing the proof.

b) Let $x \in \bar{B} - B$. That x is in the closure of B, but not B.

For every $y \in B$, select, by our assumption that X is Hausdorff, open sets 0_y and $0_{y,x}$

$$x \in 0_{y,x}, y \in 0_y$$

$$0_{y,x} \cap 0_y = \phi.$$

The sets 0_y cover B, as each point in B belongs to at least one. Since B is compact, a finite number of $0_y \cap B$ cover, and hence some

$$0_{y_1}, \cdots, 0_{y_n}$$

cover B. That is

$$B \subseteq 0_{y_1} \cup \cdots \cup 0_{y_n}.$$

Define $0 = 0_{y_1,x} \cap \cdots \cap 0_{y_n,x}$. 0 is an open set which contains x. As it does not meet any $0_{y_1}, \cdots, 0_{y_n}$, because the 0_{y_i} and $0_{y_i,x}$ are disjoint, it doesn't meet B at all.

We have produced an open set containing x, which does not meet B. Thus, $x \notin B^b$, that is x does not lie in the frontier or boundary of B. By assumption, x does not lie in B. By Definition 3.6,

$$x \notin B^i \cup B^b = \bar{B}.$$

Which contradicts our assumption, $x \in \bar{B} - B$.

In other words, $\bar{B} - B = \phi$, or $B = \bar{B}$, or B is closed.

Basic Lemma 4.2

If X and Y are compact topological spaces, then $X \times Y$ is also compact.

Proof. Let $\{0_\alpha\}$ be an open cover of $X \times Y$. Each 0_α is a union of sets of the form $V_{\alpha,x} \times W_{\alpha,y}$ where $V_{\alpha,x}$ is an open set in X and $W_{\alpha,y}$ is an open set in Y, and $(x, y) \in V_{\alpha,x} \times W_{\alpha,y} \subseteq 0_\alpha$. In other words, we take all pairs (v, w) with v running in an open set of X and w in Y. The fact that 0_α is a union of such sets is precisely the fact that these sets $V_{\alpha,x} \times W_{\alpha,y}$ are a basis for the topology on $X \times Y$. (Check this with Definition 3.13!)

It clearly suffices to show that a finite number of $V_{\alpha,x} \times W_{\alpha,y}$ sets cover $X \times Y$, for then a finite number of the 0_α's will suffice, as they are bigger.

Now, fix $x_0 \in X$, for each $y \in Y$ select $V_\alpha \times W_\alpha$ containing (x_0, y). These W_α's obviously cover Y, and as the set of pairs (x_0, y) is homeomorphic to Y and thus compact, a finite number of W_α's suffice to cover all (x_0, y) for any y in Y. This is illustrated below

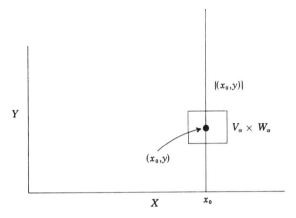

The intersection of all the V_α's, corresponding to this finite number of W_α's, is an open set V^* containing x_0. Also, note that every

$$V^* \times W_\alpha,$$

where W_α is one of these finite number of sets, lives in $V_\alpha \times W_\alpha$, and forms a basis set of $X \times Y$.

We repeat this for every $x_0 \in X$. Thus, for every $x_0 \in X$, there are a finite number of W_α's, $W_{\alpha_1}, \cdots, W_{\alpha_k}$ with each

$$V^* \times W_{\alpha_i}$$

contained in a basic set on $X \times Y$ (here V^* some open set containing x_0). This done, we turn to the first coordinate.

Select a finite number of these V^*, say $V_{\beta_1}, \cdots, V_{\beta_l}$, covering X. I claim that the sets $V_{\beta_i} \times W_{\alpha_j}$, where α's go along with the V_{β_i}'s, which were called V^*'s above, yield an open cover by a finite number of basis sets. For let $(p, q) \in X \times Y$. p belongs to some V_{β_j}. The V_{β_j} is associated with sets W_{α_i}, so that the sets

$$V_{\beta_j} \times W_{\alpha_i}$$

(i varies through a finite collection) cover $V_{\beta_j} \times Y$. Thus, q must belong to some W_{α_m}, which means

$$(p, q) \in V_{\beta_j} \times W_{\alpha_m}$$

completing the proof. This lemma will be generalized later when we prove Tychonoff's famous theorem on products of any number of spaces.

We shall now tackle the characterization of compact sets in $\mathbb{R}^n$. We shall show first that closed boxes are compact, by the help of Lemma 4.2. Then Lemma 4.1 will give the theorem. (We formerly wrote $\mathbb{R}^n$ as E^n.)

Theorem 4.1

$A \subseteq R^n$ is compact, if and only if
1) A is closed
2) A is bounded. (That is, there is a k so that if $x \in A$, $d(0, x) < k$, with 0 being the origin.)

Proof. If A is compact, we know from Lemma 4.1 that A is closed. To get A is bounded, we select a cover of A consisting of the open balls of radius 1 about each point of A. As a finite subcollection of m balls will cover, and the triangle inequality shows that any ball can be no further from 0 than the distance of its center from 0, plus 1, then if there are only a finite number of balls, and hence centers, the largest distance from a point of A to 0 is the maximum distance of a center of such a ball plus 1. Hence, A is bounded.

Intuitively, being bounded means no string of points can go off to infinity. Obviously, if a set lies in a finite number of balls of radius 1, no string of points can go off to infinity.

The converse takes more work. Let us first show that a closed interval of real numbers is compact. Let $\{0_\alpha\}$ be a collection of intersections, of open intervals, with $[a, b] \subseteq R$. That is the 0_α are basis sets for the relative topology on $[a, b]$, or equivalently basis sets for $[a, b]$ as a topological space with the usual metric topology. It suffices to show that if the 0_α cover $[a, b]$, then a finite number of them also cover $[a, b]$.

Set

$$S = \{c \mid a \leq c \leq b, [a, c] \text{ is covered by finitely many of the } 0_\alpha\}$$

S contains at least one point, namely a, as $[a, a]$ belongs to some single set 0_α. S is bounded, as all its points lie in $[a, b]$. Let d be the least upper bound.

Note that $d < b$ is impossible, for, with this d, one easily sees that $[a, d]$ must be covered by finitely many 0_α's. But then, as d belongs to some 0_β, taking this 0_β in the finitely many 0_α's, we have a finite cover which would extend to the right of d (since any open interval containing d, also has points to the right and left, or, if you prefer, bigger and smaller than d). Hence, $d < b$ implies d is not an upper bound for S.

We conclude $d = b$ and $[a, b]$ is covered by finitely many 0_α's. (Check details!)

Now, our Lemma 4.2 implies by induction that any finite product of closed intervals is compact. We note that the spaces R^n and $R \times \cdots \times R$ (n times) are obviously homeomorphic, they both consisting of n-tuples of real numbers, and the maps which take an n-tuple to itself being easily checked to be continuous and inverse to one another. It follows that any

set
$$\{(x_1, \cdots, x_n) \mid (x_1, \cdots, x_n) \in \mathbb{R}^n, a_1 \leq x_1 \leq b_1, a_2 \leq x_2 \leq b_2, \cdots,\}$$
$$a_n \leq x_n \leq b_n$$

is compact, because it is homeomorphic to a product of closed intervals. Such a set may be loosely referred to as a closed box.

Let $A \subseteq \mathbb{R}^n$ be bounded. Then, one sees at once that the x_1 coordinates of points in A are bounded, and ditto the other coordinates. Thus, there are $k_1, \cdots, k_n$ with A contained in

$$B = \{(x_1, \cdots, x_n) \mid (x_1, \cdots, x_n) \in \mathbb{R}^n, -k_1 \leq x_1 \leq k_1, \cdots, -k_n \leq x_n \leq k_n\}.$$

Intuitively, A is contained in a big box B.

But we then easily check that A is a closed set in a compact space B, so Lemma 4.1 gives the final result that A is compact.

Remarks. 1) That $[a, b]$ is compact is the famous Heine-Borel theorem.

2) Many famous sets are included under this theorem. For example, the $(n-1)$-sphere

$$S^{n-1} = \{X \mid X \in \mathbb{R}^n, d(X, 0) = 1\}$$

is obviously bounded, easily checked to be closed, and thus compact.

Theorem 4.1 is an elegant and useful criterion for compactness in $\mathbb{R}^n$. But it does not cover compactness in more general metric spaces, where examples show that our two conditions do not suffice to ensure compactness. Although we must deviate a bit from our basic results, it will be useful to some students to study some general condition for compactness. For that reason, we introduce two other concepts and prove that they characterize compactness for metric spaces. The first of these is well known in analysis.

Definition 4.3

A metric space has the *Bolzano Weierstrass property*, if any infinite subset has an accumulation point.

Recall that y is an accumulation point for $\{x_\alpha\}$, if whenever $y \in O$, O an open set, O also contains some x_α, distinct from y (if y were an x_α). The property is named for the two mathematicians who first recognized its importance for closed intervals.

Definition 4.4

A space X is *countably compact*, if when we have a countable cover $\{O_\alpha\}$, that is cover by countably many open sets, one can find a finite number of sets from the $\{O_\alpha\}$ which also cover.

64 Topology

Clearly this definition differs from compactness only in the hypothesis that one considers only countable collections of O_α's which cover.

Theorem 4.2

Let X be a metric space. Then the following three conditions are equivalent (that is each implies any other).
a) X is compact
b) X is countably compact
c) X has the Bolzano-Weierstrass property.

Proof. We shall show that a) implies c) which in turn implies b). Finally, we show that b) implies a), though we shall need an intermediate step to get this final implication. Note that we will then have a complete circle of implications

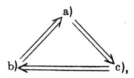

so that obviously each statement will imply any other.

We begin with a) $\Rightarrow$ c). Let $\{x_\alpha\}$ be an infinite set in X. Assume $\{x_\alpha\}$ has no accumulation point. Then, for any $y \in X$, there must be an open set O_y, $y \in O_y$, and O_y contains no point of $\{x_\alpha\}$, save possibly y itself, if it happens to be an x_α. The O_y's clearly cover (each point belongs to one), so there are a finite number of y_i's with

$$X = O_{y_1} \cup \cdots \cup O_{y_n}.$$

Now, each O_{y_i} contains at most one x_α, so there can be no more than n x_α's, contrary to the assumption that $\{x_\alpha\}$ is an infinite set. Hence, c) must be true.

Next, we show c) $\Rightarrow$ b). Assume c) and let $O_1, O_2, \cdots, O_n, \cdots$ be a countable cover of X by open sets. We suppose this is not redundant, in that no O_n is wholly contained in $O_1 \cup \cdots \cup O_{n-1}$ (if it was, we could forget it and still have a cover).

For each n, select $x_n \in O_n - (O_1 \cup \cdots \cup O_{n-1})$. Let x be an accumulation point for the x_n. Then in an open set about x, there is some x_n, not x itself. And as $x \in X$ and the O_i's cover, $x \in O_m$, for some m.

I claim infinitely many x_i's are in O_m. For if there are only finitely many, put d to be the minimum distance of x to one of these finitely many x_i's in O_m. Then

$$O_m \cap B_{d/2}(x)$$

is an open set about x which contains no x_i, except possibly x. This would contradict the definition of an accumulation point.

In particular, if there are infinitely many x_i's in 0_m, there is some x_i, $i > m$, $x_i \in 0_m$, which contradicts the way in which the x_i were constructed. We conclude that after a certain number of steps,

$$0_r - (0_1 \cup \cdots \cup 0_{r-1}) = \phi;$$

as

$$X = \bigcup_{i=1}^{\infty} 0_i,$$

and we have already removed any redundant 0_i's, we must have

$$X = 0_1 \cup \cdots \cup 0_{r-1}$$

(check!), completing the step.

Finally, we come to b) $\Rightarrow$ a). I must first detour, however, to show that if X is countably compact, there is a countable basis for the topology on X, that is a countable family of open sets, such that any open set is a union of them.

Fix $1/k$, where k is a positive integer. I claim that there are a finite number of points whose distance from one another is at least $1/k$, and any other point is nearer to one, of this finite number of points, than $1/k$. This takes a little work.

First observe that there cannot be infinitely many points no two of which are closer than $1/k$. For if there were, select a countable subfamily $x_1, x_2, \cdots$ and consider the sets

$$B_{1/k}(x_1), B_{1/k}(x_2), \cdots$$
$$B = X - \overline{(\bigcup_i B_{1/2k}(x_i))}.$$

I claim that the $B_{1/k}(x_j)$ and this B give a countable cover of X. For suppose $x \in X - B$, that is

$$x \in \overline{\bigcup_i B_{1/2k}(x_i)}.$$

Since $B_{1/2k}(x_i) \subseteq B_{1/k}(x_i)$, for all i, we may assume that for all i,

$$x \notin B_{1/2k}(x_i),$$

for otherwise x would already be covered. Hence, we may assume

$$x \in (\bigcup_i B_{1/2k}(x_i))',$$

that is x is an accumulation point of the union; then given any open set containing x, it must meet one $B_{1/2k}(x_i)$, at least. Suppose $B_{1/3k}(x)$ meets $B_{1/2k}(x_j)$, that is

$$z \in B_{1/3k}(x) \cap B_{1/2k}(x_j).$$

We then calculate

$$d(x,x_j) \le d(x,z) + d(z,x_j) < \frac{1}{3k} + \frac{1}{2k} < \frac{1}{k}.$$

That is $x \in B_{1/k}(x_j)$. This shows that the sets

$$B_{1/k}(x_1), B_{1/k}(x_2), \cdots, B$$

cover X.

By our assumption of countable compactness, there is a finite subcover. As no x_i belongs to B, infinitely many must belong to one

$$B_{1/k}(x_i)$$

which means that two points were closer than $1/k$. Thus, there are not infinitely many points no two of which are closer than $1/k$, as I had claimed.

We may now choose a selection of points, no two of which are closer than $1/k$. This must be a finite process, so eventually we will come to a point where there is no longer any point at a distance greater than or equal to $1/k$. In other words, this will then be a collection so that any other point is nearer one of them than $1/k$. Denote such a finite sequence by $x_1, \cdots, x_n$.

For each x_i in $x_1, \cdots, x_n$, consider the countable family of sets

$$0_{k,m,i} = B_{1/m}(x_i),$$

with m a positive integer. That is each $0_{k,m,i}$ is the ball of radius $1/m$ about the ith point in a collection of points, having the property that any point is nearer to one of them by at least $1/k$.

Choosing suitable x_i's for all k, we consider the countable family of open sets $0_{k,m,i}$ for all positive integers k and m and suitable x_i's. (Check that this is a countable family!)

I claim this is a basis, i.e. any open set is a union of these. This amounts to proving the claim that for any open 0, $z \in 0$, there is some $0_{k,m,i}$ with

$$z \in 0_{k,m,i} \subseteq 0.$$

For then 0 is the union of all such $0_{k,n,i}$.

Let $z \in 0$; as 0 is open, there is $\rho > 0$ with

$$z \in B_\rho(z) \subseteq 0.$$

Select k and m with

$$0 < \frac{1}{k} < \frac{1}{m} < \rho/2.$$

Then there is some x_i, associated with this choice of k, so that

$$d(z,x_i) < \frac{1}{k}$$

consider $0_{k,m,i} = B_{1/m}(x_i)$.
If $y \in 0_{k,m,i}$, then

$$d(z,y) \leq d(z, x_i) + d(x_i, y) < \frac{1}{k} + \frac{1}{m} < \rho.$$

Thus, $0_{k,m,i} \subseteq B_\rho(z) \subseteq 0$. But $d(z, x_i) < 1/k < 1/m$, so $z \in 0_{k,m,i}$, completing the proof that we have a countable basis.

To complete the step b) $\Rightarrow$ a), and thus the entire theorem, let $\{0_\alpha\}$ be a family of open sets. For each $y \in 0_\alpha$, select a set from the countable basis, say $0_{k,m,i}$ with

$$y \in 0_{k,m,i} \subseteq 0_\alpha.$$

If we could prove that a finite number of these $0_{k,m,i}$ cover X, surely this finite number of the bigger sets 0_α cover X.

But since the $0_{k,m,i}$ are countable in number, the hypothesis b) shows at once that a finite number will cover, completing the theorem.

Remark. Theorem 4.2 is substantially harder than the more basic Theorem 4.1 This is not an unreasonable price to pay, since the first only characterizes compact subsets of the $\mathbb{R}^n$ spaces, while the latter characterizes all compact metric spaces. At a minimum, the reader should be certain that he/she fully understands the first of these theorems and the statement of Theorem 4.2, before plowing on.

Problems

1. Find all spaces, which when endowed with the indiscrete topology (the minimum possible number of open sets) are Hausdorff spaces.
2. Find all topologies on the set of three points, which are Hausdorff. (It is a considerably more complicated problem to enumerate all the topologies on the set of three points.)
3. Let X denote the real line with the topology of finite complements, i.e. a set is open if and only if its complement is finite.
   Prove X is not Hausdorff.
4. Deduce Theorem 4.1 from Theorem 4.2 (*Hint*: Show that in a closed bounded set in $\mathbb{R}^n$, any infinite set has an accumulation point. This can be done by considering individual coordinates. Or, one can generalize the classical theorem, in $\mathbb{R}$, which repeatedly divides the interval in half, and chooses successively the smaller intervals which contain infinitely many points of the set in question).
5. Prove that an intersection of compact sets is compact. (Assume they all lie in a fixed Hausdorff space X.)

6. Find a non-Hausdorff space, with a non-closed compact subset.
7. Let X and Y be topological spaces and

$$f: X \to Y$$

a continuous map which is onto. Suppose X is compact. Prove that Y must then be compact. (*Hint:* Let $\{0_\alpha\}$ cover Y. Consider the cover $\{f^{-1}(0_\alpha)\}$ of X. Note that $f(f^{-1}(A)) = A$.)

8. Let $f: X \to \mathbb{R}$ be a continuous function, with X compact. Prove that f is bounded (e.g. there is K with $|f(x)| < K$ for all $x \in X$).

9. Find an example of $f: X \to Y$, a continuous onto map of topological spaces, with Y compact but X not.

10. Let $f: X \to Y$ be a continuous map, not necessarily onto, but with X compact and Y Hausdorff.

    Show that if $C \subseteq X$ is closed,

$$f(C) = \{y \mid y \in Y, y = f(x) \text{ for some } x \in C\}$$

is also closed.

11. Let $f: X \to Y$ be a continuous map between Hausdorff spaces. If X is compact and f is both 1-1 and onto, prove f is a homeomorphism.

12. a) Show that in $\mathbb{R}^n$ the closure of a bounded set is bounded.
    b) Let $f: \mathbb{R}^n \to \mathbb{R}^m$ be a continuous map. Show that the image, under f, of a bounded set is bounded.

We shall now prove the famous theorem, due to Tychonoff, on products of compact spaces. This is a strong generalization of Basic Lemma 4.2.

Theorem 4.3

Let $\{X_\beta\}$ be any family of spaces, each of which is compact. Then

$$\underset{\beta \in B}{\times} X_\beta$$

is compact.

Proof. We shall use a modification of Proposition 4.3, to the effect that X is compact, if and only if whenever $\{S_\alpha\}$ is a family of subsets (not necessarily closed), so that every intersection of a finite number of sets is non-empty, then there is a point which belongs to the closure of each S_α.

To see that this condition is indeed equivalent to the condition in Proposition 4.3, suppose the condition holds. If $\{C_\alpha\}$ is an family of closed sets, all of whose finite intersections are non-empty, then the condition asserts that

$$\underset{\alpha}{\cap} \bar{C}_\alpha \neq \phi.$$

But as C_α is closed, $C_\alpha = \bar{C}_\alpha$, and thus by Proposition 4.3 we know X is compact.

Conversely, if the condition of Proposition 4.3 holds, we apply it to the sets $\{\bar{S}_\alpha\}$. If finite intersections of S_α's are non-empty, then the finite intersections of the bigger $\bar{S}_\alpha$'s are non-empty. But then we know from the condition of Proposition 4.3 that all the $\bar{S}_\alpha$ have a point in common, showing that our new condition is also satisfied.

The reason for this new condition is that we will construct, in the course of the proof, certain sets which will not necessarily be closed. Yet we still wish to use them to show our space is compact. Onward now with the proof.

Let $\{S_\alpha\}$ be a family of sets in

$$\underset{\beta \in B}{\times} X_\beta$$

with the finite intersection property, that is finite intersections are not empty. We introduce a partial ordering on such families by

$$\{S_\alpha\} \leq \{T_{\alpha'}\}$$

whenever every S_α is actually one of the T_α's. We wish to apply Zorn's lemma (see Chapter 1); I claim that any ordered set of such $\{S_\alpha\}$ has a supremum. For if

$$\cdots \leq \{S^0_{\alpha_0}\} \leq \{S^1_{\alpha_1}\} \leq \{S^2_{\alpha_2}\} \leq \cdots$$

represents such an ordered family of such sets, I claim that the totality of all of them is also such a family. For given any finite number of sets in the total collection of all these sets, these finite number of sets must belong to some specific $\{S^k_{\alpha_k}\}$ (in fact, we can take k to be the maximum index corresponding to any one of these finite number of sets). But the $\{S^k_{\alpha_k}\}$ have the property that finite intersections are non-empty. Hence, the total collection of all the sets has the desired property; hence the ordered set of these families has a supremum, as the total collection is clearly greater than or equal to each one.

We invoke Zorn's lemma, for all such families which are greater than our given $\{S_\alpha\}$ and conclude that there is a maximal family of sets in our product space so that

1) All the original $\{S_\alpha\}$ belong to the family.

2) The sets in the maximal family have the property that all finite intersections are non-empty.

3) Any set, which meets all the sets of the maximal family, must belong to it.

Denote the maximal family by $\{M_\gamma\}$. If we can find a point which belongs to $\bar{M}_\gamma$, for all γ, we are done, for then such a point clearly belongs to every $\bar{S}_\alpha$ (recall that every S_α is some M_γ, since $\{M_\gamma\}$ is a maximal family greater than or equal to our original $\{S_\alpha\}$).

70   Topology

For any fixed $\beta_0 \in B$, define
$$\pi_{\beta_0}: \underset{\beta \in B}{\times} X_\beta \to X_{\beta_0}$$
by
$$\pi_{\beta_0}(f) = f(\beta_0)$$
for $f \in \times_{\beta \in B} X_\beta$. This is the projection on the β_0 coordinate.

One easily sees (check this) that any intersection of a finite number of the sets
$$\pi_{\beta_0}(M_\gamma)$$
which all live in X_{β_0}, must be non-empty.

Hence, the intersection of all of the
$$\overline{\pi_{\beta_0}(M_\gamma)}$$
is non-empty (this is because X_{β_0} is assumed compact). Choose, for every β_0, a point
$$u_{\beta_0} \in \bigcap_\gamma \overline{\pi_{\beta_0}(M_\gamma)}.$$
I claim that $g \in \times_{\beta \in B} X_\beta$, defined by
$$g(\beta_0) = u_{\beta_0}$$
for all $\beta_0 \in B$ is our desired point in
$$\bigcap_\gamma \bar{M}_\gamma.$$
For let
$$0 = \{f \mid f \in \underset{\beta \in B}{\times} X_\beta, f(\beta_i) \in 0_i\}$$
be a subbasis set for the topology on our product (see Proposition 3.7), with $g \in 0$. For β_i,
$$g(\beta_i) = u_{\beta_i} \in \bigcap_\gamma \overline{\pi_{\beta_i}(M_\gamma)}$$
so that 0_i must meet $\pi_{\beta_i}(M_\gamma)$ for every γ.

For any β not β_i, the image of 0 under π_β is all of X_β (this amounts to saying that in a subbasis set, a given coordinate varies in a specified open set, while all the others vary in the whole space in question). Thus,
$$\pi_\beta(0) \cap (\bigcap_\gamma \overline{\pi_\beta(M_\gamma)}) \neq \phi,$$
and thus clearly $\pi_\beta(0)$ meets all $\pi_\beta(M_\gamma)$ non-trivially, any β. (Check!)

We conclude that any such 0, $g \in 0$, meets every set M_γ in every coordinate, and one also checks at once, that 0 meets every M_γ. By maximality, 0 is some M_γ. But then any basis set, which is a finite intersection of such 0's, will also necessarily meet every M_γ.

Finally, we note that this means that g lies in the closure of every M_γ, or

$$\bigcap_\gamma \bar{M}_\gamma \neq \phi$$

which is what we set out to prove.

Remark. This theorem is, as we have remarked, a generalization of Basic Lemma 4.2, which treats the case of a product of two sets. It is unfortunate that one cannot draw pictures to illustrate this general case. Other proofs are available, for example, in the Bibliography, one may consult the books by S. Gaal or N. Bourbaki. The first gives a harder proof, but which gives a useful study of the subbasis of the product topology. The second gives a very short, slick proof, but which invokes new concepts of filters and ultrafilters. I have attempted a compromise posture, rather like the proof in the book by J. Kelley.

In any case, the theorem isn't easy, and I recommend, at a minimum, a thorough understanding of Lemma 4.2 and the statement of this theorem.

Problems

1. Show that the real numbers, as a set, are in 1-1 correspondence with a product of finite sets. (*Hint:* Let each set be the set 0, 1, $\cdots$, 9 and use the decimals to set up the correspondence.)

    Show also that the real numbers, as a topological space, are never homeomorphic to a product of spaces, each of which is finite.
2. Use Tychonoff's theorem to build an example of a compact space, whose topology does not have a countable basis.
3. Using the ideas in the proof of Theorem 4.2, find any example of a compact, Hausdorff space which is not a metric space. (*Hint:* Let $I = [0, 1]$, and form a product

    $$Z = \times I$$

    with a non-countable number of factors. Show that Z has no countable basis.)

Our final topic of the chapter is that of connectedness. We call a space disconnected, if it is the union of two disjoint open pieces (formal definition below). Otherwise, it is connected. Many basic properties, including

Topology

the intermediate value theorem of calculus, are directly involved with connectedness, and the number of connected pieces which make up a space is an algebraic invariant; that is, a number which is the same for two spaces which are homeomorphic. Some of these aspects are explored in the problems at the end of the section. Note finally that the notion of connectedness was essentially used in Problem 10, at the end of Chapter 3.

Definition 4.5

A topological space X is *disconnected*, if there are open subsets $A \subseteq X$ and $B \subseteq X$ with

$$A \neq \phi, \quad B \neq \phi$$
$$A \cup B = X$$
$$A \cap B = \phi.$$

A space is *connected*, if it is not disconnected.

As a simple example, a single point space is surely connected. The rational numbers, with the subspace or relative topology, as a subset of the real numbers, are disconnected. For let

$$A = \{x \mid x \in Q, x < \sqrt{2}\}$$
$$B = \{x \mid x \in Q, x > \sqrt{2}\}.$$

We use Q for the rational numbers here. Both A and B are clearly the intersections of Q with the open sets $\langle -\infty, \sqrt{2} \rangle$ and $\langle \sqrt{2}, \infty \rangle$ in R. Hence, they are open in the relative topology. They are obviously disjoint.

On the other hand, we have the important

Proposition 4.4

The real numbers are a connected topological space.

Proof: Assume R is disconnected, with $R = A \cup B$, $A \cap B = \phi$, $A \neq \phi$, $B \neq \phi$, A and B open. Select an interval $[a, b]$ which contains both some point in A and some point in B. (Check that this can be done if R is disconnected.)

Suppose $b \in B$, the other case being handled by the same sort of argument. Put

$$S = \{x \mid x \in [a, b], x \in A\}.$$

The points of S being bounded, we set m to be the supremum of points in S (that is the least upper bound). We shall show that m cannot lie in either A or B, which will give a contradiction.

Suppose $m \in A$. As A is open, there must be points, of an open interval, containing m, which lie entirely in A. But then there are points of A, which lie in $[a, b]$, which are bigger than m. This contradicts the assumption that m is an upper bound for the points of S, which are the points of A that lie in $[a, b]$.

Suppose $m \in B$. As B is open, there is a small open interval about m, say $\langle \alpha, \beta \rangle$ which contains only points of B. Then, the points of A, within $[a, b]$, must lie in $[a, \alpha]$. If we put

$$m' = \frac{\alpha + m}{2}$$

we have

$$\alpha < m' < m$$

so m' is a smaller upper bound for S. This contradicts the assumption that m was the *least* upper bound for S, completing the proof.

There are a variety of other useful results, which enable us to detect when a space is connected.

Proposition 4.5

Let $f: X \to Y$ be a continuous map of the topological space X *onto* the space Y. If X is connected, then so is Y.

Proof. If Y is disconnected, with $Y = A \cup B, A \cap B = \phi, A \neq \phi, B \neq \phi$, A and B open, then (check these)

$$X = f^{-1}(A) \cup f^{-1}(B),$$
$$\phi = f^{-1}(A) \cap f^{-1}(B), \quad \phi \neq f^{-1}(A), \quad \phi \neq f^{-1}(B), \quad \text{and}$$
$$f^{-1}(A) \text{ and } f^{-1}(B) \text{ are open,}$$

which would contradict the two assumptions that X is connected and f is continuous.

Proposition 4.6

If X and Y are connected topological spaces, so is $X \times Y$.

Proof. Suppose $X \times Y = A \cup B, A \cap B = \phi$, A and B open and nonempty. Choose a point $x_0 \in X$ so that

$$x_0 \times Y \subseteq X \times Y$$

meets both A and B. (Here $x_0 \times Y = \{(x_0, y) \mid y \in Y\}$.) (If no such x_0 exists, divide the points of X according as $x_0 \times Y$ is in A or in B, thus proving that X is disconnected.)

Finally, call
$$A' = \{y \mid y \in Y, (x_0, y) \in A\}$$
$$B' = \{y \mid y \in Y, (x_0, y) \in B\}$$
note $A' = A \cap (x_0 \times Y)$, $B' = B \cap (x_0 \times Y)$. One checks at once that these A' and B' show that $x_0 \times Y$ is disconnected in the relative topology as a subspace of $X \times Y$. (The set A' is not to be confused with the derived set of A, etc.)

But Y and $x_0 \times Y$ are homeomorphic (check), so that $x_0 \times Y$ could not be disconnected. This shows that our hypothesis was incorrect and $X \times Y$ is connected.

Proposition 4.7

Let X be a space, U and V two subspaces (with relative topology). Suppose $U \cap V$ is non-empty, and U and V are each connected.
Then $U \cup V$ is connected.
Proof. If $U \cup V = A \cup B$, $A \cap B = \phi$, A and B are open, non-empty subsets, one selects
$$x \in U \cap V.$$

Suppose $x \in A$. Choose $y \in B$. Note that $x \in U$ and that if $y \in U$, then $A \cap U$ and $B \cap U$ are both non-empty, as the first contains x, the second y. But these are non-empty disjoint, open sets whose union is U, which contradicts the assumption that U is connected.

Thus $y \in V$. Now $x \in U \cap V \cap A$, $y \in V \cap B$. Thus, $V \cap A$ and $V \cap B$ are non-empty. As above they are open (in relative topology) disjoint, non-empty subsets of V, whose union is V. This contradicts the assumption that V is connected.

If $x \in B$, repeat the same argument with the names changed.

Proposition 4.8

Let $Z \subseteq X$ be a subspace of X, which is connected (relative topology). Then $\bar{Z}$, the closure of Z, is also connected.
Proof. Let $\bar{Z} = (\bar{Z} \cap A) \cup (\bar{Z} \cap B)$ where A and B are open in X (that is $\bar{Z}$ is expressed as the union of two open sets in the relative topology).
Let's assume
$$\phi = (\bar{Z} \cap A) \cap (\bar{Z} \cap B) = \bar{Z} \cap A \cap B$$
and that both are non-empty.

Note that $(Z \cap A)$ and $(Z \cap B)$ are two open sets (relative topology) in Z, whose union must be all of Z and which must be disjoint. Since Z is assumed connected, one must be empty, say $Z \cap A = \phi$.

We have $Z \cap A = \phi$, $\bar{Z} \cap A \neq \phi$. Choose $u \in \bar{Z} \cap A$. Then $u \in \bar{Z}$ and $u \in A$ which is an open set in X. As A does not meet Z, u is not in the boundary Z^b.

But $\bar{Z} = Z \cup Z^b$, in fact $\bar{Z} = Z^i \cup Z^b$, so $\bar{Z} \subseteq Z \cup Z^b$, because $Z^i \subseteq Z$, and also $Z \cup Z^b \subseteq \bar{Z}$ (check). We can only conclude that $u \in Z$.

As $u \in A$, we have $Z \cap A \neq \phi$, which is a contradiction.

Remarks. All these proofs are relatively easy, but somewhat tricky. They all depend on the philosophy that to show that a space is connected one should show that it cannot possibly be disconnected. The following definition offers a possible way of showing that a space is connected directly, but it has drawbacks in that it does not cover all cases.

Definition 4.6

A space X is *arcwise connected*, if, whenever $x, y \in X$, there is a continuous function

$$f: I \to X$$

(here $I = [0, 1] = \{x \mid x \in \mathbb{R}, 0 \leq x \leq 1\}$ with

$$f(0) = x, \quad f(1) = y.$$

Some authors call this *pathwise connected*. (For most reasonable spaces, all the different variants agree here.)

Proposition 4.9

If X is arcwise connected, X is connected.

Proof. If not, select the usual A, B with $X = A \cup B$, etc.

Choose $x \in A$, $y \in B$. Choose f as in Definition 4.6, $f^{-1}(A)$ and $f^{-1}(B)$ are disjoint, non-empty, open sets in I. Hence, I is disconnected.

But in the proof of Proposition 4.4 we showed that any closed interval of real numbers is connected. Hence, the assumption that X was disconnected is false.

Corollary

A convex subset X of $\mathbb{R}^n$ (see remarks following Proposition 2.5) is connected.

Proof. Reparameterizing the line between two points to have length 1 we can always define a function $f: I \to X$ as in Definition 4.6.

We remark once again that Problem 10 at the end of Chapter 3 uses these

Topology

ideas of connectedness to show that R and R^2, the line and the plane, are not homeomorphic.

We note in addition that this is a special case of the deeper theorem on invariance of dimension, from algebraic topology, which asserts that, if R^n is homeomorphic to R^m, we must have $m = n$. (See the texts on algebraic topology in the Bibliography.)

Problems

1. Prove that X is disconnected, if and only if there is a continuous function from X to the space with two points and discrete topology (all subsets open).
2. Use Proposition 4.5 to show that if f is a continuous real valued function, defined for $x_1 \leq x \leq x_2$, and if $f(x_1) \leq c \leq f(x_2)$, then there is x_3 with $x_1 \leq x_3 \leq x_2$ so that

$$c = f(x_3).$$

This is the "intermediate value theorem."
3. Let $I = [0, 1]$. Let $f: I \to I$ be continuous. Prove that there is $x \in I$ with $f(x) = x$. (*Hint:* Use the previous problem.) This is an elementary "fixed-point theorem."
4. Show that, if $X_1, \cdots, X_n$ are finite sets, all with the discrete topology,

$$X_1 \times \cdots \times X_n$$

has the discrete topology and is thus disconnected.
5. Show that an intersection of connected subspaces need not be connected.
6. Show that the n-sphere

$$S^n = \{x \mid x \in R^{n+1}, d(x, 0) = 1\}$$

is connected for $n \geq 1$.
7. Let Z be a connected subspace of X. Suppose W is a subspace, between Z and its closure, that is

$$Z \subseteq W \subseteq \bar{Z}.$$

Prove W is connected. This generalizes Proposition 4.8.
8. Find all connected closed subsets of R.

CHAPTER 5

Sequences, Countability, Separability, and Metrization

We propose to study here the role of certain assumptions of countability (i.e. whether certain things are countable or denumerable in number). These assumptions are fruitful in many areas. For one, we learn the role of sequences in general questions about topological spaces. But a much more extensive and beautiful application is to Urysohn's general theorem on when a topological space is a metric space (that is when there is a metric, whose topology is the same as the original topology). Along the way, we will encounter another property of separation, called normality, which, while not as useful as that of Hausdorff, does figure in many beautiful and important results.

Definition 5.1

a) A space X satisfies the *first axiom of countability*, if for $x \in X$, there is a countable family of open sets, $\{0_n\}$, with x belonging to every 0_n, so that whenever 0 is an open set, $x \in 0$, there is some 0_n with
$$x \in 0_n \subseteq 0.$$

b) A space X satisfies the *second axiom of countability*, if there is a countable basis, i.e. there is a countable family of open sets $\{U_n\}$, so that any open set is a union of some of the U_n's.

Roughly the first axiom says that the topology at each point is given by countably many open sets, while the second says that the entire topology is given by countably many open sets. Of course, the total number of open sets is usually not countable.

Remarks and Examples. 1) If a space satisfies the second axiom of countability, it clearly satisfies the first axiom of countability, because if $x \in 0$, and if 0 is the union of a countable collection of open sets, then x belongs to one of these open sets, say 0_m, so that $x \in 0_m \subseteq 0$. The desired countable family is then all such 0_m which contain x.

2) There are many examples of spaces satisfying the second axiom of countability. The most important example is R. A countable basis for the open sets is the collection of all open intervals, whose diameter is rational, and whose center point is a rational number. To show that these are a basis, it is quite sufficient to show that given any open interval $\langle \alpha, \beta \rangle$, then $\langle \alpha, \beta \rangle$ is a union of such sets. But for every $x \in \langle \alpha, \beta \rangle$, we can find a (perhaps small) interval, with rational center and diameter, say $\langle r_1, r_2 \rangle$, so that

$$x \in \langle r_1, r_2 \rangle \subseteq \langle \alpha, \beta \rangle.$$

The union of all such $\langle r_1, r_2 \rangle$ clearly lies in $\langle \alpha, \beta \rangle$, as each lies in $\langle \alpha, \beta \rangle$. But as every point $x \in \langle \alpha, \beta \rangle$ belongs to at least one $\langle r_1, r_2 \rangle$, the union of all of them must be all of $\langle \alpha, \beta \rangle$.

3) A finite product of spaces satisfying the second axiom of countability clearly satisfies the second axiom also.

Thus, R^n satisfies the second axiom.

4) A subspace of a space which satisfies the second axiom of countability, endowed with the relative topology, clearly must also satisfy the second axiom of countability.

5) A space can satisfy the first axiom, but not the second. For example, a union of a non-countable collection of non-empty spaces which are disjoint but each of which satisfies the first axiom, will not satisfy the second axiom.

6) In the course of proving Theorem 4.2, we showed that a compact metric space satisfies the second axiom of countability. Clearly, a non-countable product of copies of a compact, non-trivial space, is *not* a metric space.

A somewhat more restrictive notion is as follows:

Definition 5.2

A space X is called *separable*, if there is a countable subset $S \subseteq X$ whose closure is X.

A subset, in general, whose closure is the entire space is called *dense*.

For example, if $X = R$, we may choose $S = Q$, the rational numbers.

An easy relation between these concepts is now the following:

Proposition 5.1

If X satisfies the second axiom of countability, X is separable.

Proof. Select a point in each open set 0_n of a countable basis. Call these

x_n. I claim that $\overline{\{x_n\}} = X$. For let $u \in X$, $u \notin \overline{\{x_n\}}$, that is u lies outside of the closure of the x_n's. Then there is an open set 0, containing u, with 0 not meeting any x_n. But 0 is a union of 0_n's, so for example, we may find some k,

$$u \in 0_k \subseteq 0.$$

Then $x_k \in 0_k \subseteq 0$, by construction, contradicting the assumption that 0 doesn't meet any x_n.

Proposition 5.2

A metric space satisfies the first axiom of countability.
Proof. Let $u \in 0 \subseteq X$, 0 open in the metric space X. By definition, $\exists\, \delta > 0$ so that

$$u \in B_\delta(u) \subseteq 0.$$

It is then clear that the $B_\delta(u)$'s have the desired property about each point, and hence the countable family $B_{1/n}(u) = \{y \mid y \in X, d(u,y) < 1/n\}$ are the desired countable family of open sets which contain u.

Proposition 5.3

A separable metric space satisfies the second axiom of countability.
Proof. Let X be the space and $\{x_n\}$ the countable dense subset. Consider the countable family of open balls.

$$B_{p/q}(x_n)$$

for all positive integers p, q, and n. I claim this is a countable basis.

Let $u \in X$. Select a sequence of x_n's, say x_{n_i} whose distances from u approach zero.

Suppose $u \in 0$, 0 open. Then there is a $\delta > 0$, with

$$u \in B_\delta(u) \subseteq 0.$$

Choose x_{n_i} closer to u than $\delta/3$ and p and q so that

$$0 < \frac{\delta}{3} < \frac{p}{q} < \frac{2\delta}{3}.$$

I contend that $u \in B_{p/q}(x_{n_i}) \subseteq 0$.
First,

$$d(u, x_{n_i}) < \frac{\delta}{3} < \frac{p}{q}.$$

So thus u belongs to $B_{p/q}(x_{n_i})$. Second, if $y \in B_{p/q}(x_{n_i})$

$$d(y, x_{n_i}) < \frac{p}{q} < \frac{2\delta}{3}$$

$$d(u, x_{n_i}) < \frac{\delta}{3}$$

so that

$$d(u, y) < \frac{\delta}{3} + \frac{2\delta}{3} = \delta$$

or $y \in B_\delta(u)$. As this ball lies entirely in 0, our claim is fully substantiated.

We have shown that for every $u \in 0$, there is a set of the countably family $B_{p/q}(x_n)$ which contains u and lives in 0. As before, clearly every open set is then a union of such sets.

This proposition, particularly its proof, hints at a more important role for sequences. A sequence is merely a set of elements in 1-1 correspondence with the natural numbers (or positive integers). A sequence x_n *converges* to y, if given any open set 0, with $y \in 0$, there is N so that if $n > N$, $x_n \in 0$. This is, of course, the obvious generalization of the usual notion of convergence from analysis. The following theorem shows that sequences frequently can be used to determine the topology of a space.

Theorem 5.1

Let X satisfy the first axiom of countability. Then $C \subseteq X$ is a closed subset, if and only if whenever x_n is a sequence which converges to y, and every $x_n \in C$, then $y \in C$.

Proof. Suppose C is closed and x_n is such a sequence. Suppose $y \notin C$. The definition of a convergent sequence then makes y an accumulation point for the x_n's. But as the x_n's all belong to C, y is an accumulation point for C. But C is closed, so by Proposition 3.3, $y \in C$.

Conversely suppose the condition. If C is not closed, by Proposition 3.3 there is an accumulation point for C, say v, $v \notin C$. Let 0_n be the countable family of open sets which contain v, which we assume are decreasing in size. (To do this, replace each set by the intersection of it and all the finite number of sets which come before it in the countable ordering of the 0_n. It is trivial to check that this new decreasing family still satisfies the condition of the first axiom.)

Select an x_m in each of these decreasing sets. I claim the x_m converge to v. For if $v \in 0$, 0 open, there is some set in the countable family, containing v, which lies in 0. The x_m from that set (recall we took a x_m from each set) lies in 0. As the sets are decreasing, that is, successive ones are contained in earlier ones, all the later x_m's belong to 0. We have thus found some number so that all successive x_m's belong to 0.

In other words, these x_m's converge to v. But then our condition implies $v \in C$. Thus every accumulation point of C must lie in C, or in other words, C is closed.

Corollary

Let X satisfy the first axiom of countability. Let $f: X \to Y$ be any map from the space X to the space Y. Then f is continuous, if and only if whenever x_n converges to y, then $f(x_n)$ converges to $f(y)$.

Proof. Suppose f is continuous. Then we know that whenever $0 \subseteq Y$ is open, $f^{-1}(0) \subseteq X$ is open. Suppose x_n converges to y. We need to show $f(x_n)$ converges to $f(y)$.

Let 0 be an open set in Y, containing $f(y)$. Then $f^{-1}(0)$ is an open set in X containing y. Since the x_n converge to y, there is N so that if $n > N$, $x_n \in f^{-1}(0)$. But then $f(x_n) \in 0$, proving that the $f(x_n)$ converges to $f(y)$.

Conversely, assume our condition. Suppose $C \subseteq Y$ is closed. We need show $f^{-1}(C)$ is closed. (Compare Problem 5 following Definition 3.14.) Let x_n converge to y, $x_n \in f^{-1}(C)$. By our assumption, $f(x_n)$ converges to $f(y)$. As $f(x_n) \in C$, $f(y) \in C$. Then $y \in f^{-1}(C)$.

We have shown, for the set $f^{-1}(C)$, the condition, which by Theorem 5.1 guarantees that $f^{-1}(C)$ is closed. Hence, whenever C is closed, we have $f^{-1}(C)$ is closed, which is equivalent to f being continuous (check!).

Remarks. The meaning of this theorem and corollary is that whenever we deal with spaces satisfying the first axiom of countability, we need focus our attention only on sequences to resolve questions of closedness or continuity.

Naturally, this encompasses metric spaces and many other examples. There have been various inventions designed to play the role of a sequence in arbitrary topological spaces. The nets in Moore-Smith convergence (see Kelley's book) or the filters (see Bourbaki's books) are sorts of generalized sequences and offer an effective way of studying when a set is closed in a general space. They do not, however, play any role in the basic material of this text.

Before proceeding, we need another concept.

Definition 5.3

A Hausdorff space X is called *normal*, if whenever C_1 and C_2 are two closed subsets, $C_1 \cap C_2 = \phi$, then there are open 0_1 and 0_2 with

$$C_1 \subseteq 0_1; \quad C_2 \subseteq 0_2$$
$$0_1 \cap 0_2 = \phi.$$

82   Topology

In fact, some authors do not ask that such a space be Hausdorff, but there is little sense here in studying sapces which satisfy this condition but not Hausdorff's condition (in such a space, some single point might fail to be closed).

We defer examples of non-normal spaces to the exercises at the end (such examples are not trivial), but we show here that a very broad family of topological spaces does satisfy this condition.

Proposition 5.4

Every metric space is normal.

Proof. Let C_1, C_2 be disjoint closed subsets of the metric space X, given with a metric d. For each $x \in C_1$ we define the distance from x to C_2 as

$$d(x, C_2) = \text{g.l.b.} \; d(x, y).$$
$$y \in C_2$$

That is the distance from x to C_2 is the greatest lower bound (g.l.b.) of all possible distances from x to any point in C_2. This, being a g.l.b. of non-negative numbers, must be non-negative.

But in fact, if $x \notin C_2$, $d(x, C_2)$ is positive. For if $d(x, C_2) = 0$, select a sequence of y_n so that $d(x, y_n) < 1/n$. The y_n thus converge to x. As C_2 is closed, that would mean $x \in C_2$ (check!).

This all enables us to define for $x \in C_1$

$$B_{1,x} = \{y \mid y \in X, d(x, y) < \tfrac{1}{3} d(x, C_2)\}$$
$$0_1 = \bigcup_{x \in C_1} B_{1,x}.$$

Similarly, we put for $y \in C_2$

$$B_{2,y} = \{x \mid x \in X, d(y, x) < \tfrac{1}{3} d(y, C_1)\}$$

and

$$0_2 = \bigcup_{y \in C_2} B_{2,y}.$$

Clearly, these 0_1 and 0_2 are disjoint, for if $z \in 0_1 \cap 0_2$, there are x and y with

$$z \in B_{1,x} \cap B_{2,y}$$

then

$$d(z, x) < \tfrac{1}{3} d(x, C_2) \quad \text{and} \quad d(z, y) < \tfrac{1}{3} d(y, C_1).$$

We have
$$d(x, y) \leq d(x, z) + d(z, y) < \tfrac{1}{3}d(x, C_2) + \tfrac{1}{3}d(y, C_1)$$
$$\leq \tfrac{1}{3}d(x, y) + \tfrac{1}{3}d(x, y)$$
which is absurd, proving that $0_1 \cap 0_2 = \phi$.

To complete the proof we need only show 0_1 and 0_2 are open. For this, it suffices that $B_{1,x}$ (or $B_{2,y}$) be open, which is immediate from the definition of an open set in a metric space.

In addition, the class of normal space includes all compact Hausdorff spaces, i.e.

Proposition 5.5

Every compact Hausdorff space is normal.

Proof. Let C_1, C_2 be disjoint closed sets in the compact Hausdorff space X. For each $x \in C_1$, we find first an open $0_{1,x}$, containing x, which does not meet an open set $0_{2,x,C_2}$ containing C_2. This goes as follows:

For each fixed $x \in C_1$, and any $y \in C_2$, select disjoint open $U_{1,x,y}$ and $U_{2,y}$, containing x and y, respectively. A finite number of $U_{2,y}$ is cover C_2, as a closed subset C_2 must be compact, so
$$C_2 \subseteq U_{2,y_1} \cup \cdots \cup U_{2,y_m}.$$
Set $0_{1,x} = U_{1,x,y_1} \cap \cdots \cap U_{1,x,y_m}$ which is clearly an open set about x, and which clearly does not meet our finite union of U_{2,y_i}'s, which we name $0_{2,x,C_2}$, i.e.
$$0_{2,x,C_2} = U_{2,y_1} \cup \cdots \cup U_{2,y_m}.$$
Naturally $0_{2,x,C_2}$ is an open set around C_2, as required.

The rest of the proof is remarkably similar. Choose a finite number of $0_{1,x}$'s which cover C_1, i.e.
$$C_1 \subseteq 0_{1,x_1} \cup \cdots \cup 0_{1,x_k} = 0_1.$$
Let $0_2 = 0_{2,x_1,C_2} \cap \cdots \cap 0_{2,x_k,C_2}$, that is the intersection of all the $0_{2,x,C_2}$'s for this finite family. It is then immediate that
$$C_2 \subseteq 0_2$$
and 0_2 is open. As each $0_{1,x_i}$ is disjoint from $0_{2,x_i,C_2}$, clearly
$$0_1 \cap 0_2 = \phi$$
completing the proof.

The following is a vague sketch of the proof, and I hope it is helpful, for motivation.

84 Topology

Step 1

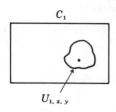
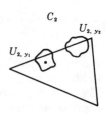

Take intersection here.   Take union here.

Step 2

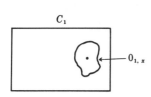
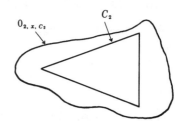

Take union here.   Take intersection here.

Problems

1. Let X_a, $a \in R$, be a family of spaces which are all homeomorphic to $I = [0, 1]$.
   Show that $\times_a X_a$ satisfies neither axiom of countability.
2. In the previous example, show $\times_a X_a$ is compact and normal, but not a metric space. Is there a subset, which is not closed, but satisfies the condition of Theorem 5.1.?
3. Show that a subspace of a metric space always satisfies the first axiom of countability and is normal. (This is very easy. Don't look for anything hard.)
4. Let $f: X \to Y$ be a continuous map from X onto Y. If X satisfies one of the countability axioms, or is separable or is normal, does the same necessarily hold for Y? (This problem will most likely seem difficult. In fact, the examples to show that the same does not hold for Y, when that is the case, are all easy. As a hint, any space which has the discrete topology is normal and satisfies the first axiom of countability.)

   The following problems are designed to show that there are Hausdorff spaces which are not normal.
5. Let X be the upper half-plane, i.e.
$$X = \{(x, y) \mid x \in R, y \in R, y \geqq 0\}.$$

Consider the collection of all closed discs which lie entirely in the set X, i.e. the sets (for $\epsilon > 0$)
$$\{(x, y) \mid (x, y) \in X, (x - x_1)^2 + (y - y_1)^2 \leq \epsilon, \epsilon \leq y_1\}.$$
Define $\sigma \subseteq X$ to be open, if whenever $(x_2, y_2) \in \sigma$, $y_2 > 0$, there is an entire disc with center (x_2, y_2) lying in σ, and whenever $(x_2, 0) \in \sigma$, there is a disc with $\epsilon = y_1$ which contains $(x_2, 0)$ as its lowest point and which lies in σ.

Prove that the set of these 0's defines a Hausdorff topology on X.

6. Let $A_1 = \{(x, 0) \mid x \text{ rational}\}$, $A_2 = \{(x, 0) \mid x \text{ irrational}\}$. Prove A_1 and A_2 are closed, disjoint subsets of X.

7. Suppose there are open 0_1 and 0_2, $0_1 \cap 0_2 = \phi$, $A_1 \subseteq 0_1$, $A_2 \subseteq 0_2$. For each $x \in A_2$, select $\epsilon_x > 0$ such that a disc of such a radius, i.e. ϵ_x, lies entirely in 0_2 and contains $(x, 0)$.

Define $B_k = \{x_1 \mid x_1 \text{ irrational}, \epsilon_x \geq 1/k, x = (x_1, 0)\}$.

Prove that there is a closed interval I on the real line, i.e. the set $\{(x, 0)\}$, so that, for some k, every subinterval contains points of B_k. (*Hint:* Suppose for each interval I and each m there is a closed subinterval J which does not meet B_m. Successively chose J_1, J_2, $\cdots$ decreasing so as J_k does not meet B_k and does not meet the k-th rational number in I, in any given order. Show that there is a point in
$$\bigcap_i J_i$$
which is not rational, and hence belongs to some B_k. Deduce a contradiction.)

8. Show that I above has a rational accumulation point for B_k. (This is trivial, as I has rational points and every interval meets B_k.)

Using this rational accumulation point, select a disc in 0_1 containing this rational point.

Prove that 0_1 and 0_2 must meet near this point. (Draw a picture.)

Deduce that X could not possibly be normal.

9. As a final problem on normal spaces, let $Y \subseteq X$ be a closed subset of a normal space X. Prove that Y is normal (Refer to Problem 3 in the set of problems which follow Definition 3.14).

The importance of normal spaces is perhaps best judged by the richness in the continuous functions from such spaces to the real numbers. The following proposition is of paramount importance here. It is due to the Russian mathematician Urysohn, and often called his lemma.

Proposition 5.6 (Urysohn's Lemma)

Let X be normal and C_1 and C_2 two disjoint, closed subsets of X. Then there is a continuous function
$$f: X \to \mathbb{R}$$

with

1) $f(x) = 0$, if $x \in C_1$
2) $f(x) = 1$, if $x \in C_2$
3) $0 \leq f(x) \leq 1$, for all $x \in X$.

Proof. Before we can give the proof, we need check a step, which I would call a lemma if it were not that someone would surely object to speaking of a lemma to a lemma. This step is needed to assure f is continuous.

Suppose $D = \{\alpha\}$ is a countable dense subset of $I = [0, 1]$ and $\{0_\alpha\}$ is a family of open sets in X with

1) $\bar{0}_\alpha \subseteq 0_{\alpha'}$ whenever $\alpha < \alpha'$
2) $\bigcup_\alpha 0_\alpha = X$.

Define $f(x) = $ g.l.b. $\{\beta \mid \beta \in D, x \in 0_\beta\}$. Then I claim $f(x)$ is continuous. A subbase of I is given by the intervals $[0, \alpha\rangle$, $\langle\beta, 1]$ defined as

$$[0, \alpha\rangle = \{x \mid 0 \leq x < \alpha\}$$

etc.

It suffices to check that the inverse image of such an interval is open. Consider the set of $y \in X$ with

$$0 \leq f(y) < \alpha,$$

and let γ run through the numbers of D less than α. I claim $f^{-1}([0, \alpha\rangle)$ is the union of the sets 0_γ. It is clear that

$$\bigcup_{\gamma < \alpha} 0_\gamma \subseteq f^{-1}[0, \alpha\rangle.$$

But if $0 \leq f(y) < \alpha$, by the definition of g.l.b., there is a γ with $f(y) < \gamma < \alpha$ and $y \in 0_\gamma$. Hence, the reverse inclusion holds, so $f^{-1}[0, \alpha\rangle$ is a union of open sets, and thus open.

On the other hand, look at the set $f^{-1}(\langle\beta, 1])$. That is the set of $y \in X$ with

$$\beta < f(y) \leq 1.$$

Because D is dense, there is $\gamma \in D$ with $\beta < \gamma < f(y)$. As $f(y)$ is the g.l.b., $y \notin 0_\gamma$, and hence, if $\delta < \gamma$, $y \notin \bar{0}_\delta$ (because $\bar{0}_\delta \subseteq 0_\gamma$). That is, $y \in X - \bar{0}_\delta$. This shows that

$$f^{-1}(\langle\beta, 1]) \subseteq \bigcup_{\beta < \delta < 1} X - \bar{0}_\delta.$$

Conversely, if $y \in X - \bar{0}_\delta$, or $y \notin \bar{0}_\delta$ for $\beta < \delta < 1$, then the g.l.b. of the indices u with $y \in 0_u$ must be as big as δ. Thus, $\beta < f(y) \leq 1$ proving the reverse inclusion. Once again, $f^{-1}(\langle\beta, 1])$, being the union of open sets, is open.

Sequences, Countability, Separability, Metrization

This completes the preliminaries. We now define a family of such open sets as follows:

By definition there are disjoint open sets about C_1 and C_2; say U_1 and U_2. Put $0_0 = U_1$, $0_1 = X - C_2$.

That is, we are invoking the above procedure for the subset $X - C_2$; $\bar{0}_0$ is a closed set which does not meet C_2 (check!) so that $\bar{0}_0 \cap C_2 = \phi$. We can choose, by normality, open sets V_1 and V_2, which are disjoint and contain $\bar{0}_0$ and C_2.

Put

$$0_{1/2} = V_1.$$

Observe that $\bar{0}_{1/2}$ does not meet V_2 and hence does not meet C_2, which is smaller.

We naturally wish to define $0_{n/2^i}$ for all positive integers n and i, so that the closure of a smaller one lies in any bigger one. We have done this for 0, $\frac{1}{2}$ and 1. Suppose we have done this for all m and 2^{i_0-1}, that is $0_{m/2^i}$ is defined for any $i \leq i_0 - 1$, and all m so that $0 \leq m/2^i \leq 1$. We wish to define

$$0_{m/2^{i_0}},$$

assuming $m/2^{i_0}$ is in lowest terms. But then m is odd, so $m - 1$ and $m + 1$ are even. Hence $(m - 1)/2^{i_0}$ is not in lowest terms, nor is $(m + 1)/2^{i_0}$, from which we conclude that the corresponding 0's are already defined and

$$0_{m-1/2^{i_0}} \subseteq \bar{0}_{m-1/2^{i_0}} \subseteq 0_{m+1/2^{i_0}}.$$

Choose open W_1 and W_2, containing $\bar{0}_{m-1/2^{i_0}}$ and $X - 0_{m+1/2^{i_0}}$, and disjoint.

Put $0_{m/2^{i_0}} = W_1$.

As W_1 does not meet W_2, which contains the complement of $0_{m+1/2^{i_0}}$,

$$\bar{0}_{n/2^{i_0}} \subseteq 0_{m+1/2^{i_0}}$$

(check!), completing the construction.

By our earlier remarks

$$f(x) = \text{g.l.b.} \{\gamma \mid x \in 0_\gamma\}$$

is a continuous function from $X - C_2$ to $[0, 1]$. $f(x) = 0$, if $x \in C_1$.

Our proposition will be complete, if we know that the definition

$$f(x) = 1, \quad \text{if } x \in C_2$$

defines a continuous function on all of X. For this, we need only check that $f^{-1}(\langle\beta, 1])$ is open, as the proof that $f^{-1}([0, \alpha\rangle)$ is open is already achieved and does not involve points of C_2 at all. But, with our new definition of f

88 Topology

for points of C_2 as well as $X - C_2$, we obviously have

$$f^{-1}(\langle \beta, 1]) = (\bigcup_{\beta < \delta < 1} (X - C_2 - \bar{0}_\delta)) \cup C_2.$$

(Check this!) In fact, this is just the union of C_2, and the inverse image of $\langle \beta, 1]$ under an original function.

If we choose any specific such δ_0, $\beta < \delta_0 < 1$, then by our construction, we have for some W_1 and W_2

$$0_{\delta_0} \subseteq \bar{0}_{\delta_0} \subseteq W_1$$
$$C_2 \subseteq W_2$$
$$W_1 \cap W_2 = \phi.$$

But this means, of course, that $\bar{0}_\delta \cap C_2 = \phi$, or

$$(X - C_2 - \bar{0}_\delta) \cup C_2 = X - \bar{0}_\delta,$$

or

$$f^{-1}(\langle \beta, 1]) = \bigcup_{\beta < \delta < 1} (X - \bar{0}_\delta)$$

which is obviously open, completing the proof.

Urysohn's lemma gives a rich variety of functions on a normal space. This rich collection of functions enables us to show that a normal space, which satisfies the second axiom of countability, is a subset of a metric space, and thus is metric (that is, there is a metric, whose topology is the original topology). This famous theorem is also due to Urysohn.

Theorem 5.2 (Urysohn)

If X is a normal space, satisfying the second axiom of countability, then X is a metric space (i.e. there is a metric d, defined on X, and the open sets are precisely those given by Definition 2.10).

Proof. Let $\{0_n\}$ be a countable basis. For each 0_n, select a point $x_n \in 0_n$, and by Urysohn's lemma a continuous $f_n: X \to I$ such that

$$f_n(x_n) = 0$$
$$f_n(y) = 1, \quad \text{if } y \in X - 0_n.$$

(Recall that in a Hausdorff space, single points such as x_n are closed subsets.)

I claim, in fact, that we may assume that if $y \in 0_n$, $f(y) < 1$, in a normal space, satisfying the second axiom of countability. Every closed set is a countable intersection of open sets which contain it (check). If, in our proof of Urysohn's lemma, we add the assumption that each set $0_{2^i - 1/2^i}$ contains the complement of a sequence of decreasing sets about C_2, then

Sequences, Countability, Separability, Metrization

we achieve the same conclusion plus the fact that every $y \notin C_2$ belongs to some $0_{m/2^i}$, with $m/2^i < 1$. Hence $y \notin C_2$ implies $f(y) < 1$.

Our construction is now easy, but we will need to verify four things to complete the proof. Recall from Example 6 of a metric space, Chapter 2, the space H of sequences of real numbers, often called Hilbert space. Define

$$\Phi: X \to H$$

by $\Phi(x) = (f_1(x), f_2(x)/2, \cdots, f_n(x)/2^n, \cdots)$. (Check that this definition is o.k.)

Recall the notation $\Phi(A) = \{y \mid y \in H, y = \Phi(x), x \in A\}$. I claim:

A) Φ is continuous.

B) Φ is 1-1.

C) If $0 \subset X$ is open, there is an open U in H with $\Phi(0) = \Phi(X) \cap U$. That is Φ sends open sets to relative open sets.

D) X is homeomorphic with a subset of the metric space H. Thus, X is a metric space, as desired.

To prove A), suppose $z_n \to z$, that is z_n converges to z in X. (Our space is necessarily first countable, so we shall use the Corollary to Theorem 5.1.) Let $f(z) \in B_\epsilon(f(z))$, $\epsilon > 0$. Choose N_0 so that

$$\sum_{i=N_0}^{\infty} \left(\frac{1}{2^i}\right)^2 < \frac{\epsilon^2}{2}.$$

For every $i < N_0$, choose an open 0_i, $z \in 0_i$, so that whenever $y \in 0_i$

$$\left(\frac{f_i(y) - f_i(z)}{2^i}\right)^2 < \frac{\epsilon^2}{2N_0}.$$

This is possible because f_i is a continuous function from X to $\mathbb{R}$. (Check this!)

We set

$$0 = 0_1 \cap \cdots \cap 0_{N_0 - 1}.$$

Because $z_n \to z$, we may find N_1, so that if $n \geq N_1$, $z_n \in 0$.

If $n > N_1$, we calculate the square of the distance from $\Phi(z_n)$ to $\Phi(z)$

$$\sum_{i=1}^{\infty} \left(\frac{f_i(z_n) - f_i(z)}{2^i}\right)^2 \leq \sum_{i=1}^{N_0-1} \left(\frac{f_i(z_n) - f_i(z)}{2^i}\right)^2$$
$$+ \sum_{i=N_0}^{\infty} \left(\frac{f_i(z_n) - f_i(z)}{2^i}\right)^2 \leq \frac{N_0 - 1}{2N_0} \epsilon^2 + \frac{\epsilon^2}{2}$$

because the terms in the second sum are all less than $(1/2^i)^2$. (Recall $0 \leq f_i(x) \leq 1$.) This is clearly less than ϵ^2.

We conclude that if $n > N_1$, $\Phi(z_n) \in B_\epsilon(\Phi(z))$, or that $\Phi(z_n)$ converges to $\Phi(z)$. Thus Φ is continuous.

To prove B), note that if $x_0 \neq x_1$, then we choose U_0 and U_1 open, $x_0 \in U_0$, $x_1 \in U_1$, $U_0 \cap U_1 = \phi$. Choose a basis set 0_k so that

$$x_0 \in 0_k \subseteq U_0.$$

Then $f_k(x_0) < 1$ while $f_k(x_1) = 1$. It follows that $\Phi(x_0)$ and $\Phi(x_1)$ differ in the k-th coordinate, proving that Φ is 1-1.

To prove C), we must show that $\Phi(0_m)$ is open in $Z = \Phi(X)$, with the relative topology. Let $\Phi(y) \in \Phi(0_m)$, that is $y \in 0_m$. Then, there is $\epsilon_m > 0$ so that if

$$|f_m(x) - f_m(y)| < 2^m \epsilon_m, \qquad x \in X,$$

then $x \in 0_m$. (This follows from the fact that f_m is defined as above. Check!)

I claim that if for some $x \in X$, $\Phi(x) \in \Phi(X) \cap B_{\epsilon_m}(\Phi(y))$ then

$$\Phi(x) \in \Phi(0_m).$$

This is obvious, since, if $\Phi(x)$ is nearer $\Phi(y)$ than ϵ_m, we have:

$$\left(\frac{f_m(x) - f_m(y)}{2^m}\right)^2 \leq \sum_{m=1}^{\infty} \left(\frac{f_m(x) - f_m(y)}{2^m}\right)^2 < \epsilon_m^2$$

or

$$|f_m(x) - f_m(y)| \leq 2^m \epsilon_m$$

or $x \in 0_m$, because of our choice of ϵ_m. In other words, if $\Phi(y) \in \Phi(0_m)$, we have found an open set about $\Phi(y)$, in the relative topology on $Z = \Phi(X)$, which lies in $\Phi(0_m)$. Thus, every point $\Phi(y)$, in $\Phi(0_m)$, lies in an open set which is contained in $\Phi(0_m)$. Thus, $\Phi(0_m)$ is a union of open sets, and hence $\Phi(0_m)$ is itself open in the relative topology.

To prove D), note that A), B) and C) show that

$$\Phi: X \to \Phi(X)$$

is 1-1, continuous, visibly onto, and sends open sets to open sets. This last condition means that the inverse map (compare Problem 11 after Theorem 4.2)

$$\Phi^{-1}: \Phi(X) \to X,$$

defined by $\Phi(y) = $ the unique x so that $\Phi(x) = y$, is continuous (check this!); Φ and Φ^{-1} set up a homeomorphism.

We conclude that X is homeomorphic to $\Phi(X)$ with the relative topology. This latter is a subset of a metric space, and hence itself a metric space. Thus means that the previously given topology on X is actually the topology which comes from a metric (as in Definition 2.10), completing the proof.

Remark. We have shown that every normal space which satisfies the second axiom of countability is homeomorphic to a subspace of Hilbert space, H. This theorem has been given various minor improvements over the years, and one may find in the literature (see the books by S. Gaal or J. Kelley) various theorems culminating in necessary and sufficient conditions for a space to be a pseudometric space (same conditions as a metric space, except that $d(x, y)$ may be zero for $x \neq y$). While these theorems are non-trivial, they all lack the beauty and basic importance of Urysohn's theorem.

Problems

1. Show that on a normal space X, the set of all real functions (continuous) separates points, i.e. if $x_0 \neq x_1$, there is a function with $f(x_0) \neq f(x_1)$.
2. Prove that on a compact Hausdorff space, all continuous real functions are bounded, but there are still enough to separate points.
3. Find an infinite set and a topology on it, so that all continuous functions to the real numbers (i.e. real functions) are constants. (This is easy! Try for several examples.)
4. A metric space is of course, normal. Find an example of a metric space which fails to satisfy the second axiom of countability. (Find a metric on an infinite set, which gives rise to the discrete topology.)

CHAPTER 6

Quotients, Local Compactness, Tietze Extension, Complete Metrics, Baire Category

We wish to discuss several final topics in point set topology, which will prove useful later on. These include quotients, local compactness, extensions, and complete metric spaces. Complete metric spaces lie outside the proper domain of topology, but are frequently useful in topology, so that we could hardly resist including them.

Quotients are delayed until this chapter because they are somewhat more sophisticated than subspaces or products, and have some rather pathological properties (they do not preserve the Hausdorff condition, etc., without some special assumptions). They are, however, enormously useful in making certain constructions.

Definition 6.1

Let $f: X \to Y$ be a map from the set X to the set Y, which is onto. Suppose X happens to be a topological space. Then, we define a topology on Y, called the *quotient topology*, by requiring that $0 \subseteq Y$ be open, if and only if $f^{-1}(0)$ is actually an open set of X. One checks trivially that this defines a topology on Y.

Examples. 1) Let $I = [0, 1]$. Define

$$f: \mathbb{R} \to I$$

by $f(x) = x$, if $0 \leq x \leq 1$; $f(x) = 0$; if $x < 0$, and $f(x) = 1$, if $x > 1$.

Then the quotient topology on I is the usual topology. To see this, we consider open sets of the subbasis $[0, \alpha)$ and $\langle \beta, 1]$ for I.

Clearly

$$f^{-1}([0, \alpha)) = \langle -\infty, \alpha \rangle,$$
$$f^{-1}(\langle \beta, 1]) = \langle \beta, \infty \rangle,$$

so that these sets are in fact open in the quotient topology. This shows that an open set in the usual topology is also an open set in the quotient topology.

Conversely, an open set in the quotient topology is clearly a union of sets whose inverse images are open intervals $\langle \gamma, \delta \rangle$, or $\langle -\infty, \gamma \rangle$ or $\langle \delta, \infty \rangle$. Thus, $[0, \alpha)$, $\langle \beta, 1]$ and $\langle \gamma, \delta \rangle$, with $0 < \gamma < \delta < 1$, are all open in the quotient topology.

Hence, an open set in the quotient topology is also an open set in the usual topology. Thus, the two topologies agree.

2) Recall that an equivalence relation is a relation $x \sim y$ (read x equivalent to y) so that $x \sim x$; if $x \sim y$ then $y \sim x$; and if $x \sim y$, $y \sim z$, then $x \sim z$. Given an equivalence relation defined on X, define the set of equivalence classes

$$X/\sim$$

to be the set of subsets of X, where every element in each subset is equivalent to every other in that subset, but not equivalent to an element not in that subset. Such a subset is called an *equivalence class*. For example, for the set $\mathbb{R}$, let $x \sim y$ mean $x - y$ is a whole number. Then

$$f \colon \mathbb{R} \to \mathbb{R}/\sim$$

is the map which assigns to each $x \in \mathbb{R}$, the subset consisting of all elements (here numbers) equivalent to x; f is clearly onto.

$\mathbb{R}/\sim$ is in this case homeomorphic to the circle

$$S^1 = \{(x, y) \mid (x, y) \in \mathbb{R}^2, x^2 + y^2 = 1\}.$$

Define $\phi \colon \mathbb{R}/\sim \to S^1$ as follows: represent S^1 as all complex numbers $e^{i\theta}$, where θ is real. Set $\phi(x) = e^{2\pi i x}$.

Now there is something to check here; x is not well-defined, and we may add or subtract any whole number to x and still be in the same class. But $e^{2\pi i n} = 1$, for any whole number n, so ϕ is well-defined. ϕ is easily continuous, 1-1, and onto.

But, according to Problem 11, after Theorem 4.2, ϕ must then be a homeomorphism. (Check this out.)

3) Let $D^2 = \{(x, y) \mid (x, y) \in \mathbb{R}^2; x^2 + y^2 \leq 1\}$. Define an equivalence relation as follows:

(x, y) is only equivalent to itself, if $x^2 + y^2 < 1$.

All pairs (x, y), with $x^2 + y^2 = 1$ are equivalent to each other.

One may show (tedious, but not hard), that the quotient space

$$D^2/\sim$$

is homeomorphic with S^2, the unit sphere in $\mathbb{R}^3$.

Proposition 6.1

Let $\sim$ be an equivalence relation defined as follows on a normal space X: $x \sim y$ whenever both x and y belong to a fixed closed subset $C \subseteq X$. Otherwise no two distinct points are equivalent. Define $f: X \to X/\sim$ and give $X/\sim$ the quotient topology, as above.

Then $X/\sim$ is Hausdorff

Proof. Consider $\{x\}, \{y\} \in X/\sim$. If neither $\{x\}$ nor $\{y\}$ comes (under f) from points in C, one easily checks that disjoint open sets in X, containing, respectively, x and y, go into (under the map f) disjoint open sets containing $\{x\}$ and $\{y\}$ (if they don't meet C). And if X is normal, clearly these open sets may be chosen not to meet C.

Suppose $x \in C, y \notin C$. Select open sets in X, 0_x and 0_y, so that $x \in C \subseteq 0_x$, $y \in 0_y$, $0_x \cap 0_y = \phi$.

Then $f(0_x)$ and $f(0_y)$ are quickly checked to be disjoint open sets in $X/\sim$ about the appropriate points.

The case where $x \notin C, y \in C$ is the same, but for a change in names.

We remark that if C is not closed, $X/\sim$ may be non-Hausdorff, even though X has virtually all desirable properties (see Problems below). Thus, quotient spaces can be tricky.

It is useful to know when a map may be built with domain a quotient space. The following characterization is important.

Proposition 6.2

Let $f: X \to Y$ be a continuous map between topological spaces. Suppose $\sim$ is an equivalence relation defined on X. Suppose that whenever $x_1 \sim x_2$, $f(x_1) = f(x_2)$. Then there is a continuous map

$$\bar{f}: X/\sim \to Y$$

defined by $\bar{f}(\{x\}) = f(x)$. (Here, we write $\{x\}$ for the subset of elements equivalent to x.)

Proof. $\bar{f}$ is well-defined, because $f(x)$ is constant on all x in any given equivalence class. The only thing that has to be proved is that $\bar{f}$ is continuous. Let

$$\rho: X \to X/\sim$$

associate with each x, its equivalence class. $0 \subseteq X/\sim$ is open, if and only

if $\rho^{-1}(0)$ is open. Now, I claim
$$\rho^{-1}(\bar{f}^{-1}(U)) = f^{-1}(U)$$
for any $U \subseteq Y$ (check this).

Hence, if $U \subseteq Y$ is open, $\rho^{-1}(\bar{f}^{-1}(U))$ is open, because f is continuous and this is exactly $f^{-1}(U)$. We conclude that $\bar{f}^{-1}(U)$ is a set in $X/\sim$ with the property that ρ^{-1} of it is open. By definition, it must be open.

In summary, if $U \subseteq Y$ is open, $\bar{f}^{-1}(U)$ is open, or, in other words, $\bar{f}$ is continuous.

Remarks. The situation in Proposition 6.2 is often described by saying that we have a *commutative diagram*

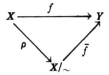

This means $f = \bar{f} \cdot \rho$. In general, a commutative diagram of maps means that, whenever two compositions have the same initial and final set (or space), the maps are equal.

To define a map from $\mathbb{R}/\sim$ in Example 2 above, one needs only a continuous map from $\mathbb{R}$, which is constant on the sets
$$\cdots \alpha - n, \quad \alpha - n + 1, \cdots, \alpha, \quad \alpha + 1, \cdots, \alpha + n_1 \cdots.$$

We now turn to a second brief, but useful, topic, that of local compactness. This is the most common of a variety of generalizations of compactness.

Definition 6.2

X is *locally compact*, if, whenever $x \in X$, there is an open 0, $x \in 0$, with $\bar{0}$ compact.

Examples. 1) $\mathbb{R}$ is locally compact (compare with Theorem 4.1).

2) Any compact space is locally compact, trivially.

3) Q, the rational numbers, is not locally compact. In fact, any closed interval of rational numbers (more than 1) contains sequences converging to an irrational. Such a sequence, in Q, is an infinite sequence without an accumulation point, in Q. (Check this out with Theorem 4.2(c), and we have that such an interval cannot be compact.)

Proposition 6.3

Let X be Hausdorff, locally compact, but not compact.
Then there is a compact Hausdorff space $\hat{X}$, and a map

$$i\colon X \to \hat{X}$$

which is 1-1, continuous, sends open sets to open sets, and whose image (that is $i(X)$) consists of all $\hat{X}$ except one point.

Proof. Let $\hat{X} = X \cup \{w\}$, where w designates a point not in X. We give X the following for open sets:
1) All open sets in X are defined to be open sets in $\hat{X}$.
2) If $K \subseteq X$ is compact,

$$(X - K) \cup \{w\}$$

is defined to be open.

The verification of the axioms for a topology is straightforward, for example, let 0_1 and 0_2 be two open sets, one of each type. Then

$$0_1 = (X - K) \cup \{w\} \quad \text{and} \quad 0_2 \subseteq X \text{ is open.}$$

Then $0_1 \cap 0_2 = [(X - K) \cup \{w\}] \cap 0_2 = (X - K) \cap 0_2$ which is open in X and hence in $\hat{X}$. The other verifications are just as easy (check).

We must prove that $\hat{X}$ is compact and Hausdorff. For this, suppose $\hat{X} = \bigcup_\alpha 0_\alpha$. Let 0_{α_0} be any one of the 0_α's which contains w. Then, there is $K \subseteq X$, compact, with $(X - K) \cup \{w\} = 0_{\alpha_0}$. The rest of the 0_α's, omitting 0_{α_0}, cover K, so choose a finite subcover

$$K \subseteq 0_{\alpha_1} \cup \cdots \cup 0_{\alpha_n}.$$

Then every point of $\hat{X}$ lies in one of the finite number of sets $0_{\alpha_0}, 0_{\alpha_1}, \cdots, 0_{\alpha_n}$ (check!).

To show $\hat{X}$ is Hausdorff, note that if $x, y \in X \subseteq \hat{X}$ since X is Hausdorff we may find the desired open sets trivially. On the other hand, consider $x \in X$ and w. Because X is locally compact, we may find an open set 0 with $x \in 0$, and $\bar{0}$ compact. Then 0 and $(X - \bar{0}) \cup \{w\}$ are disjoint open sets which contain x and w, respectively, showing that $\hat{X}$ is Hausdorff.

Finally, since X is included in $\hat{X}$ we may define $i(x) = x$, giving a 1-1 map which takes open sets to open sets by definition. We must show i is continuous. For this, it suffices to check that $i^{-1}(0)$ is open when $0 = (X - K) \cup \{w\}$, K compact (the other case being immediate). But in this case,

$$i^{-1}(0) = X - K$$

(check), which is clearly open.

Remarks. $\hat{X}$ is called the one-point compactification of X.

If X is not Hausdorff and locally compact, then $\hat{X}$ will not be Hausdorff (see Problems below).

Problems

1. Show that the torus, $S^1 \times S^1$, is homeomorphic to the quotient space
$$D/_\sim$$
where $D = \{(x, y) \mid (x, y) \in \mathbb{R}^2; 0 \leq x \leq 1, 0 \leq y \leq 1\}$ and $\sim$ is the equivalence relation

$$(x, 0) \sim (x, 1), \quad \text{all} \quad 0 \leq x \leq 1$$
$$(0, y) \sim (1, y), \quad \text{all} \quad 0 \leq y \leq 1$$

(and any (x, y), $0 < x < 1, 0 < y < 1$ is equivalent only to itself).

2. Define an equivalence relation on the set of real numbers $\mathbb{R}$ as follows:
$$x \sim y \quad \text{whenever} \quad x - y \text{ is rational.}$$
Prove that $\mathbb{R}/_\sim$ is not Hausdorff.

3. Show that for any space X and equivalence relation $\sim$, the map
$$\rho: X \to X/_\sim$$
defined by $\rho(x) = \{x\}$ (where $\{x\}$ means the set of elements in X which are equivalent to x) is a continuous map of X onto $X/_\sim$.

   Conclude that if X is compact or connected, then $X/_\sim$ is compact or connected (respectively).

4. Show that S^n is homeomorphic to the one-point compactification of E^n. (*Hint*: Take a copy of E^n which is tangent to S^n at some point x, inside E^{n+1}. Let $\tilde{x}$ be the antipode of x, that is the unique other point of S^n on the line through x and the center of the sphere. Define a map of E^n onto $S^n - \tilde{x}$ by sending any point $y \in E^n$ into the unique point where the line through y and $\tilde{x}$ meets S^n. Draw a picture in E^3.)

5. Find the one-point compactification of the half-open interval $[0, 1)$.

6. Show that a finite Cartesian product of locally compact Hausdorff spaces is a locally compact Hausdorff space. Is this true for an infinite product?

We now wish to head toward the famous Tietze extension theorem. For this, we need a bit about uniform convergence. We shall define a sequence of functions as converging uniformly, if it satisfies Cauchy's condition uniformly, that is as follows:

Definition 6.3

Let f_n be a sequence of continuous functions from X, any space to $\mathbb{R}$.

We say that it converges *uniformly* if given $\epsilon > 0$, there is an $N > 0$ so that for any n and m bigger than N, and any x whatsoever,
$$|f_n(x) - f_m(x)| < \epsilon.$$

Topology

The important thing here is that given $\epsilon > 0$, there is some N which works for all x. We now show that such a sequence approaches a limit function $f(x)$.

Proposition 6.4

Let X be a space, f_n a sequence of continuous functions to R, which converges uniformly. Then
A) For each fixed x, $f_n(x)$ converges to a number $f(x)$.
B) The function $f(x)$ is continuous.

Proof. For a fixed x, the sequence of numbers $f_n(x)$ is called a Cauchy sequence; students familiar with analysis will recognize this proof as standard.

Set $b_n(x) = $ greatest lower bound of the sequence $f_n(x), f_{n+1}(x), \cdots$. (One easily checks that such a sequence is bounded.) The $b_n(x)$ are clearly increasing. But, for any n, with $\epsilon > 0$ and N as before, we have

$$b_n(x) \leq \max\,(f_1(x), \cdots, f_N(x), f_{N+1}(x) + \epsilon).$$

(Check this from our hypothesis!)

Hence, the $b_n(x)$ have a least upper bound, which we call $f(x)$.

I claim that the $f_n(x)$ converge to $f(x)$. For we may choose N_1 so large that

$$|\,b_n(x) - f(x)\,| < \frac{\epsilon}{3}$$

whenever $n > N_1$ (the $b_n(x)$ are increasing and $f(x)$ is their least upper bound).

We may also choose N_2 so that

$$|\,f_n(x) - f_m(x)\,| < \frac{\epsilon}{3}$$

if $n, m > N_2$. As $b_n(x)$ is the greatest lower bound in the sequence $f_n(x), f_{n+1}(x), \cdots$, we may find some $f_m(x)$, with $m > n > N_2$, which is nearer to $b_n(x)$ than $\epsilon/3$.

Then if $n > N = \text{Max}\,(N_1, N_2)$,

$$|f_n(x) - f(x)| \leq |f_n(x) - f_m(x)| + |f_m(x) - b_n(x)| + |b_n(x) - f(x)|$$

where $f_m(x)$ is that element nearer to $b_n(x)$ than $\epsilon/3$, among the $f_n(x), f_{n+1}(x), \cdots$.

As each term is less than $\epsilon/3$, we have the desired result, proving the assertion A).

To prove B), we must prove that for each $\alpha < \beta$, $f^{-1}(\langle \alpha, \beta \rangle)$ is open. Let x be any point with $\alpha < f(x) < \beta$. Choose $\epsilon > 0$ so that

$$\alpha < f(x) - \epsilon < f(x) + \epsilon < \beta.$$

Choose N so that $n > N$ implies $|f_n(x) - f(x)| < \epsilon/3$, all x. (Check that this is possible! The important thing is that it work for all x.)

For some fixed $n > N$, choose an open set 0 so that if $x_1 \in 0$,

$$|f_n(x_1) - f_n(x)| < \frac{\epsilon}{3}.$$

Then, if $x_1 \in 0$,

$$|f(x) - f(x_1)| \leq |f(x) - f_n(x)| + |f_n(x) - f_n(x_1)| + |f_n(x_1) - f(x_1)|;$$

as each is less than $\epsilon/3$, it follows that

$$f(0) \subseteq \langle f(x) - \epsilon, f(x) + \epsilon \rangle.$$

$f^{-1}(\langle \alpha, \beta \rangle)$ is then the union of all such open sets, and hence is open.

This proposition is most frequently used in analysis, but the following extension theorem is another application.

Theorem 6.1 (Tietze)

Let X be a normal space, A a closed subset. Suppose we have given a continuous function

$$f: A \to I = [0, 1]$$

(of course, A has the relative or subspace topology). Then there is a continuous function

$$F: X \to I$$

so that if $x \in A$, $f(x) = F(x)$. Such a F is called an *extension* of f.

Proof. We need a trivial extension of Uryshon's lemma to the effect that if C_1 and C_2 are disjoint closed sets in X, $a < b$, there is a continuous

$$g: X \to [a, b]$$

with

$$g(x) = \begin{cases} a, & \text{if } x \in C_1 \\ b, & \text{if } x \in C_2. \end{cases}$$

(Check this. The function of Proposition 5.6 may be easily modified to make it work.)

Let $C_1 = f^{-1}([0, \frac{1}{3}])$, $C_2 = f^{-1}([\frac{2}{3}, 1])$. These are disjoint closed sets in X, so there is a continuous $g_1: X \to [\frac{1}{3}, \frac{2}{3}]$ with

$$g_1(x) = \frac{1}{3}, \quad \text{if} \quad x \in C_1$$

$$g_1(x) = \frac{2}{3}, \quad \text{if} \quad x \in C_2.$$

Topology

Clearly,
$$|f(x) - g_1(x)| \leq 1/3$$
for all $x \in A$ (check!).

Set $h_1(x) = f(x) - g_1(x)$. Put $D_1 = h_1^{-1}([-1/3, -1/3^2])$ and $D_2 = h_1^{-1}([1/3^2, 1/3])$. Once again, we may find, $g_2: X \to [-1/3^2, 1/3^2]$ with $g_2(x) = -1/3^2$, if $x \in D_1$, $g_2(x) = 1/3^2$, if $x \in D_2$, and
$$|h_1(x) - g_2(x)| \leq 2/3^2, \quad \text{for} \quad x \in A.$$
In other words,
$$|f(x) - (g_1(x) + g_2(x))| \leq 2/3^2$$
whenever $x \in A$.

Continuing in this way (check this!), we find a sequence of functions, defined on X, $g_n(x)$, with
$$|f(x) - (g_1(x) + \cdots + g_{n-1}(x))| \leq \frac{2^{n-2}}{3^{n-1}}$$
for $n \geq 2$ and $x \in A$. Also, each $g_n(x)$ has range $[-2^{n-2}/3^n, 2^{n-2}/3^n]$.

Put $F_n(x) = g_1(x) + \cdots + g_n(x)$. If $n < m$
$$|F_n(x) - F_m(x)| \leq |g_{n+1}(x)| + \cdots + |g_m(x)| \leq \frac{2^{n-1}}{3^{n+1}} + \cdots + \frac{2^{m-2}}{3^m}.$$
Clearly, the $F_n(x)$'s will then converge uniformly to a suitable function $F(x)$, as in the above proposition.

But, if $x \in A$,
$$|f(x) - F(x)| = |f(x) - (g_1(x) + \cdots + g_n(x)) - \sum_{i=n+1}^{\infty} g_i(x)|$$
$$\leq |f(x) - (g_1(x) + \cdots + g_n(x))| + |\sum_{i=n+1}^{\infty} g_i(x)|$$
$$\leq \frac{2^{n-1}}{3^n} + \sum_{i=n+1}^{\infty} \frac{2^{i-2}}{3^i}.$$

Letting n go to infinity, we can easily make this less than any $\epsilon > 0$. Hence $|f(x) - F(x)|$ is less than any preassigned positive number, so it must be 0. This completes the proof.

We shall complete this chapter by discussing complete metric spaces. The definition here is motivated by the considerations like Definition 6.3. The notion of a complete metric space is not topological in that two metric spaces, one complete, the other not, may still be homeomorphic. Still, the notion is very useful in topology and of great importance in analysis.

Definition 6.4

Let X be a metric space, $\{x_n\}$ a sequence of elements of X. Denote the distance function by d. The sequence $\{x_n\}$ is called a *Cauchy sequence*, if given $\epsilon > 0$, there is a positive integer N, so that whenever n and m are greater than N,

$$d(x_n, x_m) < \epsilon.$$

Definition 6.5

A metric space X is *complete* if every Cauchy sequence $\{x_n\}$ in X converges to an element of X (see Theorem 5.1).

Examples. 1) The real numbers with usual metric are complete. This is a well-known fact in analysis. The proof is very close to the first part of the proof of Proposition 6.4.

2) The open interval $\langle 0, 1 \rangle$, with the usual metric, i.e. $d(x_1, x_2) = |x_1 - x_2|$ is not complete. For $x_n = 1/n$ is a Cauchy sequence which converges to 0, which lies *outside* $\langle 0, 1 \rangle$. Note that R and $\langle 0, 1 \rangle$ are homeomorphic. Of course, it is possible to put a metric on $\langle 0, 1 \rangle$, which gives rise to the usual topology, but which is complete.

3) The rational numbers Q, with the usual metric, are not complete. The sequence 1, 1.4, 1.41, 1.414, $\cdots$ converging to $\sqrt{2}$ is a Cauchy sequence which does not converge to any point in Q.

4) There are many more sophisticated examples. For one, Hilbert space H (Example 6 of a metric space, in Chapter 2) is complete. This is a standard result in analysis. (See for example, E. C. Titchmarsh, *Theory of Function*, Oxford University Press. This occasionally goes by the name of the Riesz-Fisher theorem, though it is more exactly a lemma of J. von Neumann.)

Definition 6.6

A) Let $S \subseteq X$ be a subset of a space (usually, but not necessarily, a metric space). Let 0 be an open set of X. S is *not dense in* 0, if $0 \cap \bar{S}$ is a proper subset of 0.

B) S is *nowhere dense*, if S is *not dense* in any open set in X.

C) S is of the *first category*, if S is the union of a countable (or finite) family of sets S_i, each of which is nowhere dense.

D) S is of the *second category*, if it is *not* of the first category.

Remarks. This terminology is admittedly archaic, but it has stuck in the literature, and seems to defy any effort to change it. The concepts are most useful for complete metric spaces, where the notions of first and second category actually serve to distinguish characteristics which are of interest in topology. Here are some examples.

Examples. 1) In R, a single point is nowhere dense, and thus of the first category.

2) $Q \subseteq R$ is of the first category.

3) It will follow from the following theorem that any set $S \subseteq R$, which contains an open set, is of the second category.

Theorem 6.2 (Baire Category Theorem)

Let X be a complete metric space. Then any open set $0 \subseteq X$ is of the second category.

Proof. Suppose $0 = \bigcup_i S_i$, where $i \in Z, i \geq 0$, and each S_i is nowhere dense.

Let 0_1 be an open ball, of radius ρ, in 0.

Let 0_2 be an open ball, of radius less than $\rho/2$, $\bar{0}_2 \subseteq 0_1 - (\bar{S}_1 \cap 0_1)$. This may be chosen, because, due to our assumptions, $0_1 - (\bar{S}_1 \cap 0_1)$ is a non-empty open set.

Let 0_n be an open ball, of radius less than $\rho/2^{n-1}$, $\bar{0}_n \subseteq 0_{n-1} - (\bar{S}_{n-1} \cap 0_{n-1})$.

These balls are nested, in that

$$0_1 \supseteq 0_2 \supseteq 0_3 \supseteq \cdots.$$

Any two points in 0_n are closer than $\rho/2^{n-2}$. Furthermore, all these balls are non-empty, and any point $x \in 0_n$ lies outside $S_1 \cdots S_{n-1}$.

We consider the sequence of centers of these open balls, say c_n. This is clearly a Cauchy sequence because if $n, m > N$, we have $c_n, c_m \in 0_N$, so

$$d(c_n, c_m) < \rho/2^{N-1},$$

and of course, this goes to zero. Let c_n converge to c.

We shall complete the proof by showing that c can't live in 0, or outside 0.

Suppose $c \in 0$; then $c \in S_m$ for some m. But here $c_{m+1}, c_{m+2}, \cdots$ all live in 0_{m+1}, and

$$\bar{0}_{m+1} \subseteq 0_m - (\bar{S}_m \cap 0_m).$$

Since $\bar{0}_{m+1}$ is closed and $c_{n+1}, c_{m+2}, \cdots$ converge to c (check this), $c \in \bar{0}_{m+1} \subseteq 0_m$, contradicting $c \in S_m$.

Suppose $c \notin 0$. Then $c \notin \bar{0}_m$ for all $m > 1$. Hence

$$c \notin \bigcap_m \bar{0}_m.$$

But $c_m, c_{m+1}, \cdots$ all lie in 0_m, because our sequence of balls is nested. Thus, as before, $c \in \bar{0}_m$, for all m, trivially giving a contradiction. In other words, the assumption that $c \notin 0$ is also false.

As $c \in 0$ and $c \notin 0$ are both false, and $c \in X$ (X is a complete metric space and c is the limit of a sequence of points in X), we can only conclude that our original assumption, that

$$0 = \bigcup_m S_m$$

must be false. Hence, 0 is not of the first category; that is to say, 0 is of the second category.

Corollary

A complete metric space, or any subspace of it which contains an open set, must be of the second category.

Proof. If such a set is a countable union of nowhere dense sets, then any subset would be a countable union of nowhere dense sets. (Check!)

Thus, if a set containing an open set was a countable union of nowhere dense sets, the open set would have to be a union of nowhere dense sets. But that would contradict the previous theorem (the Baire category theorem).

The Baire category theorem may be used in a variety of ways. Here is a typical example.

We shall prove that the plane is not a countable union of lines. Suppose it is, i.e.

$$\mathbb{R}^2 = \bigcup_n L_n$$

where each L_n is a line in $\mathbb{R}^2$, that is the graph $Ax + By + C = 0$. I claim that each line is nowhere dense. For if 0 is an open set which meets L_k at the point (x, y) there is an open disc about (x, y) entirely within 0. But the closure of a line in this disc is just the line, and hence there are points of 0 not in this closure. We illustrate this as follows.

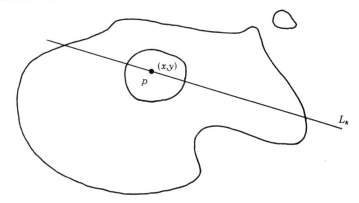

p is a point in 0 not on the closure of L_k.

But then if our assumption that R^2 is a countable union of lines is correct, R^2 would be of the first category. This would contradict the fact that R^2, being a complete metric space (check!), is of the second category.

Problems

1. Prove that if X is a normal space, A a compact subspace,
$$f: A \to R^n$$
a bounded continuous function, then f has an extension to a continuous
$$F: X \to R^n$$
(i.e. $F(x) = f(x)$, if $x \in A$). (*Hint:* The range of a bounded function lies in a product of closed intervals. Note that a function with range R^n is the same thing as an ordered collection of n functions with range R).
2. Show that Tietze's extension theorem is false if one omits the assumption that A is closed.
3. Let X be a normal space which is arcwise connected (Definition 4.6). Prove that there is a continuous function from X onto I, if and only if X has more than one point. What if X is not normal?
4. Prove that the irrational numbers, as a subset of the real numbers with usual metric, are of the second category.
5. Prove that a closed subset of a complete metric space is a complete metric space.
6. Show that the sphere
$$S^2 = \{(x, y, z) \mid (x, y, z) \in R^3, x^2 + y^2 + z^2 = 1\}$$
is not the union of a countable number of circles (these being intersections of S^2 with planes).

The following problems are somewhat more involved.

7. Show that if R is given the half-open interval topology ($[\alpha, \beta)$ is a basis set), then $R \times R$ is not normal. (*Hint:* Consider the line $x + y = 0$. Both the set of rational points, and the set of irrational points on this line are closed. Suppose they had disjoint open neighborhoods 0_1 and 0_2. Then the set of irrational points, which lie on a half-open square of sides greater than $1/n$, within the open set 0_2, is nowhere dense in this line, with the usual topology. Thus, were our supposition true, the irrational numbers would be a first category set in R.)
8. Let X be a complete metric space, 0_n a countable family of open sets, which are all dense. Prove that
$$\bigcap_n 0_n$$
is dense.

CHAPTER 7

Generalities about Manifolds; The Classification of Surfaces

We now wish to narrow our study to a very important class of spaces, the manifolds. Manifolds occur very naturally in diverse branches of mathematics, such as Lie groups, complex analysis, differential equations etc. They represent a very natural and beautiful class of spaces, and, in one sense, they form the best, global generalization of Euclidean spaces. Intuitively, they are the spaces which locally (i.e. in a neighborhood of any point) look like Euclidean space.

Manifolds are nice enough so that a good deal of mathematics, of the sort that we do in Euclidean space, can be carried over to them (with suitable assumptions). They can be classified in low dimensions (we do this at the end of the chapter). They also have the pleasant property of avoiding much of the local pathology which can make the study of arbitrary spaces—even arbitrary metric spaces—a very difficult matter.

Our goal is to do manifolds in general, and then to study some very basic examples, such as projective spaces. Finally, we close the chapter with the classification of an important class of 2-dimensional manifolds (the triangulated surfaces).

Definition 7.1

A non-empty topological space X is an n-dimensional manifold, or n-manifold for short, if
1) X is Hausdorff
2) If $x \in X$, there is an open set 0, containing x, which is homeomorphic to $\mathbb{R}^n$.

Remarks and Examples. 1) $\mathbb{R}^0$ is, as a matter of convention, a single point. Any space X, with the discrete topology, is a 0-dimensional manifold (0-manifold for short). For then we may choose the point itself as the open set containing that point.

2) R^n is an n-manifold. Choose, for 0, the entire space R^n.

3) Let $U \subseteq R^n$ be an open set. Then U is an n-manifold. For suppose $u \in U$. Since $u \in R^n$, we may choose an open ball, with center u, say B, so that
$$u \in B \subseteq U \subseteq R^n;$$
B is homeomorphic to R^n. (To check this, express a point in polar coordinates—that is a radius and various angles—and use a modified tangent function on the radius leaving the angles alone. The proof is a straightforward generalization of the fact that $\langle 0, 1 \rangle$ is homeomorphic to R.)

4) $S^1 = \{(x, y) \mid (x, y) \in R^2, x^2 + y^2 = 1\}$ is a 1-manifold. Choose ϵ with $0 < \epsilon < 1$. Let
$$U_1 = \{(x, y) \mid (x, y) \in S^1, y > -\epsilon\}$$
$$U_2 = \{(x, y) \mid (x, y) \in S^1, y < +\epsilon\}.$$
Pictorially

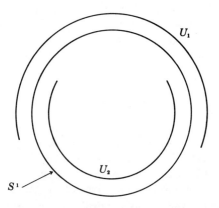

any point lies in U^1 or U^2 and each is homeomorphic to $\langle 0, 1 \rangle$, or equivalently, homeomorphic to R.

More generally, S^n is a manifold of dimension n.

5) If we drop the assumption that the space be Hausdorff, in this definition, then some rather grotesque things can occur. For example, let our space be

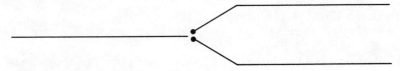

that is the open half-line $\langle -\infty, 0 \rangle$, next to which we put two closed lines $[0, \infty \rangle$. The open sets about any point, except 0, are the usual. An open set containing 0 should contain either

a) The union of $\langle \alpha, 0 \rangle$ with an interval $[0, \beta \rangle$ in the upper line ($\alpha < 0$, $\beta > 0$) or

b) The union of $\langle \alpha, 0 \rangle$ with an interval $[0, \beta \rangle$ in the lower line.

One easily checks that the Condition 2) of Definition 7.1 is satisfied, but the space isn't Hausdorff. Such examples are not reasonable generalizations of Euclidean space, so they have been traditionally excluded from the definition.

6) An open subset of a manifold is a manifold.

Manifolds avoid certain difficulties which arise for arbitrary spaces. Our first proposition shows that, if they are connected, the manifolds are always arcwise connected.

Proposition 7.1

Let M be an n-manifold. Then M is connected, if and only if M is arcwise connected.

Proof. By Proposition 4.9, we need only prove the half of the assertion which says that if M is connected, M is arcwise connected. We shall pick $x_0 \in M$ and let N denote the set of points $y \in M$ so that there is a continuous map

$$f: I \to M$$

with $f(0) = x_0$ and $f(1) = y$. N is visibly an arcwise connected subset of M. We shall show that N is both open and closed, so that if $N \neq M$, then $M - N$ is both open and closed; this would imply immediately that M is disconnected.

M visibly satisfies the first axiom of countability because locally it is the same as $\mathbb{R}^n$. Suppose $\{x_k\}$ is a sequence of points in N, $\lim x_k = z$, i.e. x_k converges to z. I claim $z \in N$. For this, choose an open nbd. of z, i.e. a set 0, which is open and contains z, and which is homeomorphic to $\mathbb{R}^n$. For large k, $x_k \in 0$. As $x_k \in N$, define

$$f: I \to M, f(0) = x_0, f(1) = x_k;$$

as x_k and z both live in 0, which is homeomorphic to $\mathbb{R}^n$ and hence arcwise connected, choose a continuous

$$g: I \to M, g(0) = x_k, g(1) = z.$$

Define

$$h: I \to M$$

by

$$h(t) = \begin{cases} f(2t), 0 \leq t \leq \frac{1}{2} \\ g(2t-1), \frac{1}{2} \leq t \leq 1. \end{cases}$$

As $f(1) = g(0) = x_k$, this defines a continuous map with $h(0) = x_0$, $h(1) = z$, as desired. Hence, $z \in N$ and N is closed.

To show N is open, let $z \in N$, and choose once again an open 0, homeomorphic to R^n, with

$$z \in 0 \subseteq M.$$

If z may be joined to x_0 by a continuous arc, i.e. a function such as h above, then since any point in 0 may be joined to z by a continuous arc, we may use the very same procedure, as in building h above, to join x_0 to an arbitrary point in 0.

Hence, $0 \subseteq N$. For every $z \in N$, we have found an open set 0 so that

$$z \in 0 \subseteq N.$$

N is visibly the union of all these 0, and hence, N is open.

The procedure of building h from f and g will be used, in a basic way, in Chapter 8.

We wish to study several further basic properties of manifolds. For reasons of convenience, we shall study these primarily in the case of compact manifolds. The results all admit generalizations, but the general theorems would lead us far afield (see H. Whitney in the Bibliography for more general results).

Definition 7.2

Let X be a topological space, $\{0_\alpha\}$ a cover of X by open sets with the property that every point x belongs to only finitely many 0_α's.

A *partition of unity*, subordinate to the cover $\{0_\alpha\}$, is a collection of functions $\{f_\alpha\}$, indexed by the same set of α's, so that

1) $f_\alpha : X \to I$
2) $f_\alpha(x) = 0$, if $x \notin 0_\alpha$.
3) $\sum_\alpha f_\alpha(x) = 1$ for all $x \in X$.

Note that by our assumption on $\{0_\alpha\}$ and Property 2), the sums occurring in Property 3) are finite sums. The condition 3) is what gives rise to the name of this concept.

Proposition 7.2

Let M be a compact n-manifold, i.e. an n-manifold which happens to be compact. Then there is an open cover of M by a finite number of sets, $\{U_\alpha\}$, homeomorphic to R^n, and a partition of unity $\{f_\alpha\}$ subordinate to that cover. (We may also assume that the sets $f_\alpha^{-1}((0, 1])$ cover M.)

Manifolds and Classification of Surfaces   109

Proof. Choose an open cover by sets, each of which is homeomorphic to $\mathbb{R}^n$, and a finite subcover

$$U_1, \cdots, U_k.$$

For each x in any U_i, choose an open set $V_{x,i}$ with

$$x \in V_{x,i} \subseteq \bar{V}_{x,i} \subseteq U_i.$$

The $V_{x,i}$'s may be taken to be homeomorphic to $\mathbb{R}^n$, for example they may be open balls around x in U_i, which is homeomorphic to $\mathbb{R}^n$.

Let

$$0_1, \cdots, 0_m$$

be a finite number of $V_{x,i}$'s which cover M (recall M is compact). In general, m will be much bigger than k.

Then each 0_i has its closure in some U_j. Use Urysohn's lemma (Proposition 5.6) to select a set of functions (m of them, but we write the index as j).

$$g_j \colon M \to I$$

with $g_j(x) = 1$, if $x \in \bar{0}_i$; $g_j(x) = 0$, if $x \notin U_j$. Since the 0_i's cover, for any x there is a j with $g_j(x) = 1$.

Thus, $\sum_j g_j(x) > 0$ for all x. Put

$$f_i(x) = \frac{g_i(x)}{(\sum_j g_j(x))}.$$

These are visibly continuous functions from M to I. The sum

$$\sum_j f_j(x)$$

is clearly 1. If $x \notin U_j$, $g_j(x) = 0$ and hence $f_j(x) = 0$. As the 0_i's cover, the last assertion is immediate. In general, we will have more than k functions.

The concept of a partition of unity is of broad usage in the topology of manifolds (not merely the compact ones). It may be used to introduce various structures in manifolds (see for example, any modern text on differential geometry). We shall apply it here to prove that a compact manifold is homeomorphic to a closed subset in some Euclidean space.

Proposition 7.3

Let M be a compact n-manifold. Then there is a 1-1, continuous function

$$\phi \colon M \to \mathbb{R}^N,$$

for some suitably large N, so that ϕ maps M homeomorphically onto $\phi(M)$, having the relative topology as a subset of R^N.

Such a ϕ is called an *imbedding* of M in R^N.

Proof. Let $\{U_i\}$ be an open cover for which we have a partition of unity $\{f_i\}$, $1 \le i \le k$, as in the previous proposition. Let $N = (n+1)k$, and we regard R^N as a Cartesian product of k copies of R^{n+1}, i.e.

$$R^N = R^{n+1} \times \cdots \times R^{n+1}$$

k-factors.

In each copy of R^{n+1}, let D^n be the closed ball of radius 1 about the origin in R^n, which is the subspace of R^{n+1}, with last coordinate zero. Let C^{n+1} denote the cone over D^n, with vertex the point $(0, 0, \cdots, 0, 1)$. This may be described pictorially as

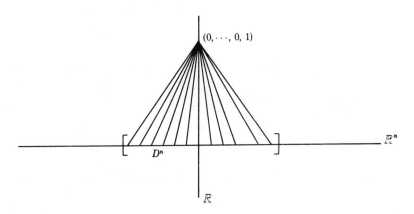

We define a function

$$\phi_i : M \to C^{n+1} \subseteq R^{n+1}$$

by

$$\phi_i(x) = \begin{cases} (f_i(x) \cdot x, 1 - f_i(x)) \in R^{n+1}, & \text{if } x \in U_i \\ (0, \cdots, 0, 1), & \text{if } x \notin U_i \end{cases}$$

where we regard U_i as the same as the interior of $D^n \subseteq R^n \subseteq R^{n+1}$. Note that this makes sense because, if $x \in U_i$, x is regarded as a point of D^n, and hence is an ordered n-tuple of real numbers. Because $f_i(x)$ approaches 0 as x runs through a sequence approaching the boundary of U_i, $\phi_i(x)$ is easily checked to be continuous. (Note that $f_i(x) \cdot x$ means the scalar product of the real number $f_i(x)$ and the vector x in R^n.)

Define
$$\phi(x) = (\phi_1(x), \cdots, \phi_k(x)) \in \mathbb{R}^{(n+1)k}.$$

ϕ is continuous (visibly), so we must show that it is 1-1 and sends open sets in M onto relatively open sets in $\phi(M)$.

Let $x, y \in M$. Suppose $x, y \in U_i$ for some fixed i. If $f_i(x) \neq f_i(y)$, then clearly $\phi(x) \neq \phi(y)$, for they differ in at least one coordinate. If $f_i(x) = f_i(y)$ and $\phi(x) = \phi(y)$, then checking the appropriate coordinates, we see that
$$f_i(x) \cdot x = f_i(y) \cdot y$$
or
$$x = y.$$

(Check that we may assume $f_i(x) \neq 0$.) On the other hand, for any x, some $f_i(x) > 0$, that is $x \in U_i$. If $y \notin U_i$, then $f_i(x) = 0$. This immediately implies that $\phi(x) \neq \phi(y)$. Hence, ϕ is 1-1.

Since M is compact, $\phi(M)$ is compact. Thus
$$\phi: M \to \phi(M)$$
is a continuous, 1-1 map from a compact Hausdorff space to another. If K is a closed subset of M, it is compact. $\phi(K)$ is then compact and, hence, closed. Thus ϕ is a 1-1 map which sends closed sets to closed sets. Because a 1-1 map sends complements to complements, ϕ is open, i.e. sends open sets to open sets.

Then, ϕ^{-1} is well defined, and continuous, and hence, ϕ is a homeomorphism. (Compare this last part of the proof with Problem 11 after Theorem 4.2.)

Corollary

A compact manifold is a metric space (in fact a subspace of $\mathbb{R}^N$).

There are many deeper such theorems (see the book by H. Whitney).

We now will examine a further class of manifolds, which are important in topology, algebraic geometry, etc. These are the projective spaces, which may be described in a variety of different ways, although all involve quotient spaces. The projective spaces are the first class of important spaces which cannot be visualized or perceived clearly in the low 3-dimensional world in which we live. Other classes of spaces, for example the spheres or the Euclidean spaces, are easy to imagine because their general features are displayed by the cases which we "see" in 2 or 3 dimensions. But the 1-dimensional projective space is the same as the circle, and the 2-dimensional projective space is sufficiently complex so as to be not homeomorphic to any subspace of $\mathbb{R}^3$ (and thus beyond our powers of visual perception).

Historically, they arose out the idea that, from a visual point of view, the plane has a circle of points infinitely far away. For example, two rails in a railbed seem to meet at infinity (visually) and the collection of all such points where parallel lines appear to meet (forward and backward being regarded as the same) appears to be a circle.

It is possible to define projective spaces in terms of ideas such as these, but we will adopt another viewpoint in the interest of getting a clear and quick exposition. Our description will show that the projective spaces are "covered" by the spheres in a 2-fold way. That is, there is a very natural, two-to-one map from the n-dimensional sphere, S^n, to the n-dimensional projective space, which will be studied in greater depth in the later chapters.

Definition 7.2

$$S^n = \{X \mid X \in \mathbb{R}^{n+1}, d(0, X) = 1\}.$$

Introduce an equivalence relation

$$X \sim Y, \quad \text{if either} \quad X = Y \quad \text{or} \quad X = -Y$$

on S^n. (Recall that if $X = (x_1, \cdots, x_{n+1})$, then $-X = (-x_1, \cdots, -x_{n+1})$.) The point $-X$ is called the *antipode* of X.

We define the n-dimensional *projective space* as

$$P^n = S^n/\sim.$$

Remarks. 1) Technically, we have defined the n-dimensional *real* projective space.

2) There is the obvious continuous map

$$\rho: S^n \to P^n$$

defined by $\rho(X) = \{X\}$, that is $\rho(X)$ is the set of elements equivalent to X (2 in number), which itself is by definition a single element of P^n. ρ is clearly onto.

Proposition 7.4

P^n is a compact, connected, n-dimensional manifold.

Proof. The map ρ is a two-to-one map of S^n onto P^n. Since S^n is compact and connected, we conclude at once that P^n is also. (See Problem 7 after Theorem 4.2 and Problem 6 after Proposition 4.9.)

To show that P^n is a manifold, let $\bar{x} \in P^n$, $\bar{x} = \rho(x)$, with $x \in S^n$. Choose a small open set O containing x, lying nearer to x than the equator in the plane perpendicular to x. Or, in other words, choose an open set, containing x, with all points in O making an angle of less than $\pi/2$ with x. We illustrate this now, the set O looking like a polar ice cap minus its edge.

Manifolds and Classification of Surfaces   113

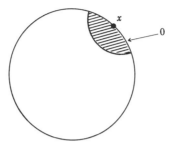

Since no two points of 0 are antipodes, ρ maps 0 in a 1-1 fashion to a set containing $\bar{x}$ in P^n. But a set in P^n is open, by the definition of the quotient topology, exactly when ρ^{-1} of the set is open. As ρ is 1-1, restricted to 0,

$$\rho^{-1}(\rho(0)) = 0 \cup \text{(antipodes of points in 0)},$$

so that $\rho(0)$ is open in P^n. The same reasoning shows that ρ maps open sets in 0 to open sets in $\rho(0)$, so we conclude that ρ is a homeomorphism of 0 to $\rho(0)$.

Hence, as 0 may be chosen to be homeomorphic to $\mathbb{R}^n$, because S^n is an n-manifold, we see that any $\bar{x} \in P^n$ lies in an open set, here denoted $\rho(0)$, which is homeomorphic to $\mathbb{R}^n$. Thus, P^n is an n-manifold.

Remarks. The projective plane P^2 is already sufficiently complex as to not be homeomorphic to a subset $\mathbb{R}^3$. One may show, using the results of the later portions of this chapter, that P^2 is homeomorphic to the quotient space of a disjoint union of a Möbius band and a disc, identifying together the points on the boundary of each (the boundary of a disc or a Möbius band is a circle). A Möbius band is the result of twisting a strip of, say, paper by 180° and gluing the ends together. It may be regarded as a subspace of $\mathbb{R}^3$.

In other words, let X be a disjoint union of a disc (circle + interior) and Möbius band. Introduce a parameter, running from 0 to 2π, on the boundary of each. Two points, one on the boundary of the disc, the other on the boundary of the Möbius band, are equivalent if they have the same values of the parameters. All other pairs of points are inequivalent. The quotient space is then (as we shall see) homeomorphic to P^2.

We now pass to a further class of examples.

Definition 7.3

A topological space X is called a *topological group*, if
1) it is a group (i.e. there is an operation, or a map $\mu: X \times X \to X$), with operations x^{-1} (the inverse) and identity $e \in X$, so that
   a) $\mu(x, \mu(y, z)) = \mu(\mu(x, y), z)$ (associative law)

b) $\mu(x, x^{-1}) = \mu(x^{-1}, x) = e$ (inverse law)
c) $\mu(x, e) = \mu(e, x) = x$ (identity law).

2) The map

$$\phi: X \times X \to X$$

defined by $\phi(x, y) = \mu(x, y^{-1})$, is continuous.

Algebra texts often write $x \cdot y$ rather than $\mu(x, y)$.

Remarks. 1) Any group, with the discrete topology, is a topological group. (For then $X \times X$ has the discrete topology and the map in question is continuous.)

2) The real numbers form a topological group with $\mu(x, y) = x + y$. Clearly $\phi(x, y) = x - y$ is continuous (in two variables).

3) The circle is a topological group. It is homeomorphic to the quotient group of $\mathbb{R}$ by the subgroup Z, the integers.

Definition 7.4

A *Lie group* is a topological group X, which is also an n-manifold for some $n > 0$. (n is called the dimension of the Lie group.)

For example, the circle S^1 is a Lie group. More generally,

Proposition 7.5

The group of $n \times n$ matrices ($n \times n$ square arrays of real numbers, with multiplication $(A \times B)_{ij} = \sum_{k=1}^{n} A_{ik}B_{kj}$), whose determinant is not zero, is a Lie group. It is denoted $GL(n; \mathbb{R})$, and has the topology of a subset of the set of all $n \times n$ matrices, which is the same as $\mathbb{R}^{n^2}$.

Proof. We shall show that $GL(n; \mathbb{R})$ is an open set in $\mathbb{R}^{n^2}$, or that

$$\{x \mid x \text{ an } n \times n \text{ matrix, } \det(x) = 0\}$$

is a closed set in $\mathbb{R}^{n^2}$. For this, it suffices to show that if

$$\lim x_n = x$$

with x_n, x all $n \times n$ matrices, and if $\det(x_n) = 0$ for all n, then $\det x = 0$. But this is an immediate consequence of the fact that the determinate is a continuous function (in fact a polynomial function) of the n^2-variables, i.e.

$$\det (\lim_n x_n) = \lim_n \det (x_n) = \lim_n 0 = 0$$

(see Corollary to Theorem 5.1).

It is obvious that the map $(A, B) \to A \times B^{-1}$ is continuous, since these operations may be described explicitly via functions of the entries. (B^{-1} may be expressed in terms of B and determinants.)

The Lie groups (named after S. Lie) are basic and important examples. The original definition was more restrictive then the one which we have given, but the efforts of many mathematicians have shown that our definition is equivalent to the original one (see, for example, the book of Montgomery and Zippin listed in the Bibliography).

We cannot resist one further definition which describes a class of manifolds which are of paramount importance in research at this time.

Definition 7.5

Let M be an n-manifold. We say that M is a *differentiable manifold*, if M is covered by open sets $\{U_\alpha\}$, each homeomorphic to R^n with a homeomorphism

$$\phi_\alpha: U_\alpha \to R^n$$

so that whenever

$$x \in U_\alpha \cap U_\beta$$

the function $\phi_\alpha \cdot \phi_\beta^{-1}$ which maps a subset of R^n to a subset of R^n, as indicated by the two maps here

$$R^n \supseteq \phi_\beta(U_\alpha \cap U_\beta) \xrightarrow{\phi_\beta^{-1}} U_\alpha \cap U_\beta \xrightarrow{\phi_\alpha} \phi_\alpha(U_\alpha \cap U_\beta) \subseteq R^n,$$

is always a differentiable map (i.e. has all partial derivatives of any order and these are continuous).

For example, R^n is clearly a differentiable manifold for we may cover R^n by one open set, and our condition is vacuously satisfied. Spheres are also examples (see the Problems below) and it is known that a Lie group is always a differentiable manifold.

Problems

1. Prove that if M_1 and M_2 are compact manifolds, then $M_1 \times M_2$ is also a compact manifold.
2. Let M be a manifold $x_0 \in M$. Define the arcwise connected component of x_0, written M_{x_0}, to be the set of $x \in M$ so that there is a continuous

$$\phi: I \to M$$

with $\phi(0) = x_0, \phi(1) = x$.
    Prove
    A) M_{x_0} is always an open set in M.
    B) M_{x_0} is a connected manifold.
    C) $M_{x_0} = M$ for every $x_0 \in M$, if and only if M is connected.
    (Compare with Proposition 7.1.)
3. Show that $GL(n; R)$, the set of $n \times n$-matrices with real entries and determinant not zero, is a disconnected manifold. Find the number of connected components (maximal disjoint, connected, open subsets).

4. Prove that the Cartesian product of two topological groups is a topological group.
5. Show that S^1 (or more generally S^n) is a differentiable manifold, i.e. find a cover by open sets, homeomorphic to $\mathbb{R}$, such that the functions $\phi_\alpha \cdot \phi_\beta^{-1}$ (see Definition 7.5) are differentiable.
6. Show that S^1 is the only compact connected 1-manifold.

We shall now focus down on 2-dimensional manifolds (often called surfaces for short). We wish to obtain a classification of these, but we shall do this under the assumption, that the manifolds be triangulated (Definition 7.6 to follow). This assumption is unnecessary in that every 2-manifold may be triangulated, but it would involve an unreasonable detour into other areas to prove this. (A good text, which also does not prove this, but contains a full list of references, is W. S. Massey *Algebraic Topology; An Introduction*.)

The main result says—to put it roughly—that any compact 2-manifold may be obtained by gluing together, in a reasonable way, spheres, toruses, and projective spaces (precise details to follow). This shall be achieved by several stages. First we shall show that every compact 2-manifold is a quotient space of a closed disc

$$D^2 = \{(x, y) \mid x^2 + y^2 \leq 1\}$$

by an equivalence relation that only identifies pairs of distinct points which lie on the boundary.

Then we shall show that such a quotient of D^2 can, by a variety of easy tricks, be put into a certain special form. Then we identify that special form with the desired conclusion.

We will detour, however, to study several special examples which serve as a model for what follows. We shall consider a disc D^2, or for short D, which is the unit circle in the plane, plus its interior. We shall study cases where the boundary, that is the circle, is divided into several line segments lined up end-to-end, and these segments are traditionally denoted with letters such as $a, b, c, \ldots$. For example

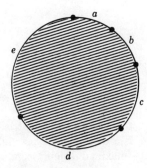

Manifolds and Classification of Surfaces

But more specifically, we are interested in quotient spaces of this space under equivalence relations which identify points on the boundary.

In the rest of this chapter, these identifications or equivalences will be of a very simple type. We shall, for example, glue the points of one segment, say a, to the points of another segment, say c. This means that we introduce a coordinate in a, which runs in a linear increasing way from 0 to 1, and one in c which also runs from 0 to 1. The point at coordinate t in a is defined to be equivalent to the point in c at time t. Other points in D, not in a or c, are defined to be equivalent only to themselves in this example. (Check that this defines an equivalence relation!) The quotient space is then referred to as the space obtained from D by identifying a with c, or, in the most loose language, we shall speak of the space, obtained from D, by gluing a to c. For simplicity, we might then call c by the name a, also.

Now—as you perhaps already noticed—my remarks in the previous section are not well-defined. The reason for this is that one can introduce a parameter in a (or equivalently c) in two possible ways, running from 0 to 1 in a clockwise or a counterclockwise direction. One can easily see that the quotient spaces can be very different with these different choices. Thus it is important to have a notation to specify a way in which we wish to parameterize the segments. Two pieces of notation are common here, and we shall use both. If we write a, we assume a clockwise orientation, that is the parameter is to increase in a clockwise direction. Otherwise we write a^{-1}. The second form of notation is to put an arrow on the segment a to indicate the direction in which the parameter is to increase.

We shall now look at various examples; I think the concepts will then be clear.

Examples. 1) *A tube, pipe, or cigar band.* Consider D

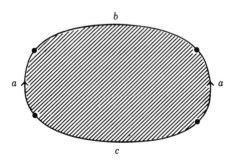

This picture means that we identify corresponding points on the two segments denoted a, forming a space

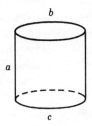

Or, in other words, we introduce an equivalence relation $\sim$ on D where two points in the different segments, denoted a, are equivalent if they are in the same vertical position, and all other pairs of different points are inequivalent. Our space is then $D/\sim$.

Traditionally, one draws such a disc as a square, which is easily seen to be homeomorphic to the disc. (Check.) That is

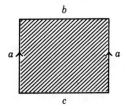

2) *The Möbius band*

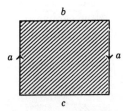

The only difference here is that the two segments here are parameterized in different directions. The quotient space, which has only one side (in the technical terminology "non-orientable"), looks as follows

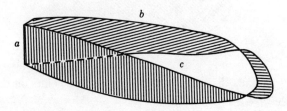

In this space, b and c become lined up end-to-end, and form a circle which is the boundary of the Möbius band. If we were interested in metrics, the distance around the boundary of the Möbius band would be the sum of the distances along b and c.

3) *The Sphere S^2*

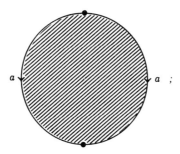

you can easily check that this quotient space, which is obtained by "sewing together" the two halves of the boundary of D in the same direction, is homeomorphic to S^2.

4) *The projective plane P^2*

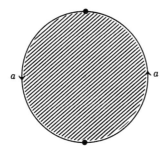

I claim that P^2 is homeomorphic to the quotient space of D, as indicated, by "sewing together" the points of the boundary of D in opposite directions. To see this, note that this quotient space of D is visibly homeomorphic to the quotient of D by the equivalence relation which identifies diametrically opposed points on the boundary. On the other hand, P^2 is the quotient of S^2 by identifying all pairs of diametrically opposed points (i.e. antipodal points) with one another.

I contend that in forming a quotient space $D_1/\sim$, it suffices to consider a subspace $D_0 \subseteq D_1$ which contains at least one point in each equivalence class. For we may define a map

$$h: D_0/\sim \to D_1/\sim$$

by sending each class in D_0 into the class to which it belongs in the bigger

D_1. Since D_0 contains at least one element in every class, h is immediately onto. h is 1-1 because if two points are equivalent in D_1, and they both lie in D_0, then they are equivalent in D_0; in fact, it is the very same equivalence relation. This map is clearly continuous, so that it must be a homeomorphism (see Problem 11 after Theorem 4.2) when D_1 is compact.

Thus, in forming P^2, we can restrict our attention to the upper (closed) hemisphere of S^2, rather than S^2 itself, because those points in the upper hemisphere contains at least one of every pair of antipodal points.

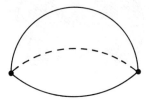

The only points which get identified are diametrically opposed points on the equator. But this is visibly homeomorphic to the quotient of D which identifies diametrically opposed points in the boundary (to check this, map the upper hemisphere onto the disc spanned by the equator by projecting downward; this is easily checked to be 1-1, continuous, and onto—hence a homeomorphism because the spaces involved are compact).

5) *The Klein bottle.* This is defined as the following "twisted" quotient of D.

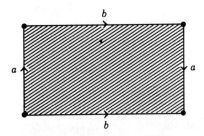

6) *The torus* (see Problem 1 after Proposition 6.3)

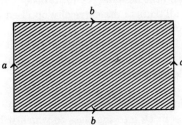

These examples should lend some credence to the idea that any compact 2-manifold can be obtained as such a quotient of a disc. But it might appear, at this stage, that these constructions are quite distinct from one another. We therefore will consider two further examples, which will throw some light on our later classification.

7) The projective plane P^2 is homeomorphic to the result of removing an open disc from S^2 and sewing to the circle, which is the boundary of the disc, the boundary of the Möbius band.

The space, which we would like to show is homeomorphic to P^2, may be notated

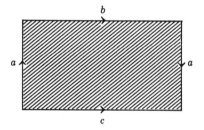 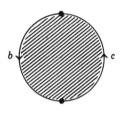

in the notation which we have been using. If we identify the two b's together in the picture, we clearly have

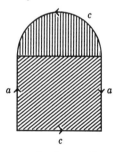

Furthermore, if we simultaneously change the orientation (direction of the parameter) of both copies of c, clearly nothing is changed. Hence, our space must be homeomorphic to

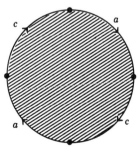

which is visibly the result of identifying diametrically opposed points on the boundary of a disc, that is to say, by 4) above, P^2.

8) The Klein bottle is homeomorphic to the result of cutting out two distinct discs from S^2 and sewing in two Möbius bands as in Example 7. We shall show that the Klein bottle is in fact the result of sewing two Möbius bands along their boundaries. From there it is easy to check that the result is the same—up to homeomorphism—to two Möbius bands sewed to a thin band, as in Example 1. Of course, the thin band is homeomorphic to a sphere with two distinct open discs removed (check this).

We represent the Klein bottle as follows.

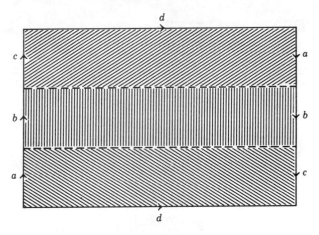

It is trivial to check that the dotted line represents a homeomorphic copy of a circle in the space, and that the region between the dotted lines is a Möbius band, identified with the circle along the dotted line.

To see that the rest of the space is also a Möbius band, consider

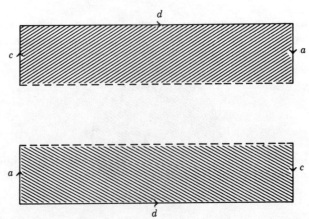

Manifolds and Classification of Surfaces 123

We can just as soon write this as

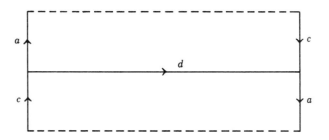

or

and this is obviously a Möbius band with the dotted line as boundary, which is what we required.

Definition 7.6

A (finite) *simplicial complex* (of dimension 2)—called a 2-complex, for short—is a quotient space of a disjoint union of points, closed intervals I and triangles (including the interior), under an equivalence relation which either

a) identifies vertices (single ends of distinct intervals or vertices of triangles) to one another, or

b) identifies different triangles by linearly gluing (or identifying) single edges together (or possibly makes no identifications).

Consult Definition 10.4 for the general definition of a finite simplicial complex.

Remarks. 1) The simplest examples would be a point or a closed interval, or a triangle. These building blocks are called *simplexes*.

2) A more fancy example is the tetrahedron (boundary only)

124   Topology

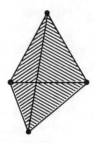

There are 4 vertices, 6 intervals or edges, and 4 triangles. The space is visibly a quotient space of 4 triangles under an equivalence relation which identifies points on edges linearly.

3) For an example of a quotient space of a line and a triangle, which is *not* a simplicial complex, consider

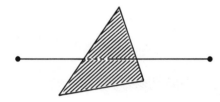

That is a quotient of an edge and a triangle which identifies together two interior points. For another example *not* a complex, consider the circle as a quotient of only 1 edge. This violates 1), because it identifies together two vertices of the same interval.

4) Clearly, the definition generalizes to n-dimensions, where an n-simplex is defined as, for example, the intersection of all closed, convex sets in $\mathbb{R}^n$ which contain $(n + 1)$ independent points. (See Bibliography for details in this topic in algebraic topology and our treatment in Chapter 10.)

A *subcomplex* of a complex is a subset which is a union of some of the simplexes in the complex.

Theorem 7.1

Let M be a connected, compact 2-dimensional manifold which is assumed to be a 2-complex. Then M is a quotient space of the closed 2-disc, i.e.

$$D = \{(x, y) \mid x^2 + y^2 \leq 1\},$$

under an equivalence relation which identifies together a finite number of pairs of closed intervals along the boundary.

Proof. Consider the collections of subsets of M, which are 2-complexes, and which are homeomorphic to the closed 2-disc D. As M is a 2-complex, with only a finite number of 2-simplexes, there are only a finite number of subcomplexes. As every single 2-simplex or triangle is homeomorphic to D, the collection of such subsets of M is non-empty.

We select a maximal such complex K, that is a maximal union of simplexes (vertices, edges, and triangles) in M, which is homeomorphic to D. Maximal means no bigger one is homeomorphic to D. The existence of such a K is clear, because there are finitely many simplexes in M.

I claim that if L is a subcomplex of M (or any complex for that matter) which is homeomorphic to D, and T is a triangle with one edge in common with L, that is one edge linearly identified with L and touching L only along that edge, then $L \cup T$ is also homeomorphic to D. This statement, which is an important step in the proof, will be proved in steps.

First, note that L is homeomorphic to D, with the edge, along which T will be glued, homeomorphic to an interval a on the boundary

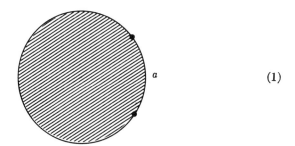

(1)

This is then homeomorphic to the half-disc

$$\{(x, y) \mid x^2 + y^2 \leq 1, x \leq 0\}$$

with a corresponding to $\{(0, y) \mid -1 \leq y \leq 1\}$.

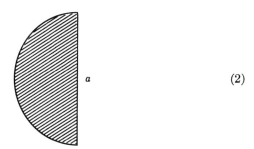

(2)

To check this, note that our figure (1) is homeomorphic to

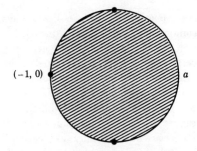

with a taking-up half the boundary. We may follow this by a homeomorphism which contracts all line segments from $(-1, 0)$, which lie in the circle, so that they only reach the y axis

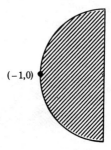

(check that this is a well-defined homeomorphism).

On the other hand, the single triangle which we wish to attach or identify to K, along the edge a, is clearly homeomorphic to the other half of the disc

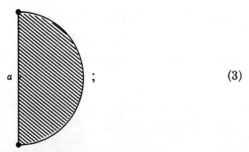 ; (3)

gluing the two together both (2) and (3) proves that the new space, the union of K and the triangle, is homeomorphic to the entire disc.

To finish the proof, let $T_1, \cdots, T_j$ be the triangles in M, not in K, the maximal subcomplex. Since M is connected, we may assume they are ordered in such a way that each one, say T_i shares a face with $K \cup T_1 \cup \cdots \cup T_{i-1}$. We add T_1 to K, but identify only one face with K. We leave the remaining faces unidentified and call the new simplex, with some identifications *not* made, S_1. Thus, we have

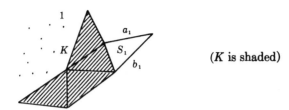

(K is shaded)

giving the names a_1 and b_1 to the faces of the new complex, which must be identified later to get $K \cup T_1$. Our previous remarks show that $K \cup S_1$ must still be homeomorphic to D.

As T_2 must meet $K \cup T_1$ in a face, we add a triangle S_2 to $K \cup S_1$, identifying points along one face but leaving the remaining faces to be identified later, i.e.

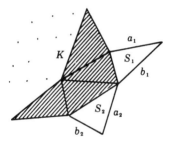

Once again $K \cup S_1 \cup S_2$ is homeomorphic to D, by an earlier argument, and after suitable identification $K \cup T_1 \cup T_2$ is a quotient of $K \cup S_1 \cup S_2$, by making the desired identification along faces. Of course, S_2 may share an edge with S_1, rather than K. (Note that it is easy to see that exactly two triangles must meet along a given edge. Since our space is a 2-dimensional manifold, one may deduce this by removing a suitable interval from a neighborhood of a point. For another proof, see Remark b), after Corollary 8 in Chapter 10.)

Continuing in this fashion, $K \cup T_1 \cdots \cup T_j$ is a quotient of

$$K \cup S_1 \cdots \cup S_j$$

which is homeomorphic to D, under linear identifications of segments of the boundary only.

But $M = K \cup T_1 \cdots \cup T_j$, so the proof is complete.

This is the only place where we make explicit use of the notion of *complex*. We will, however, manipulate various triangles in the next theorem.

Our immediate goal is now

128 Topology

Theorem 7.2

Any compact, connected 2-manifold, M, which is a complex, is homeomorphic to a 2-sphere, out of which a finite number of holes have been removed, and for which we have replaced in these holes either

a) handles, i.e. a tube (Example 1 before) is identified in a pair of holes—one end in each hole

b) a cross-cap or Möbius band, i.e. the boundary of a Möbius band is identified with the boundary of the hole (see Example 7).

Proof. From the previous theorem, we know that M is a quotient of a 2-disc, with lines of the boundary identified. It is our claim that M is equivalently a disc D with three possible types of identification:

a) Two edges which are to be identified are adjacent and are to be identified in opposite directions (we write a^{-1} as well as showing the arrow)

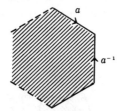

In a simple case

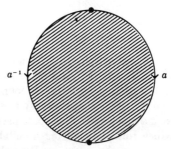

This yields the sphere S^2.

b) Two edges which are to be identified are adjacent, but in the same direction.

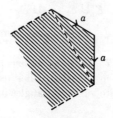

This is a cross-cap or Möbius band attached to a hole (compare Example 8). For the dotted line represents a hole and the triangle which it forms with the a's a Möbius band. To check this, take a Möbius band

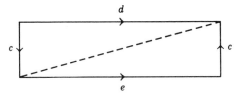

cut along the dotted line, indicating this by two a's, which are to be identified. This yields a new representation of the same space as

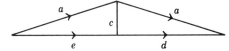

Regarding ed as the dotted line in our first figure—it is clearly a circle (check!)—we see that the claim, that the triangle is homeomorphic to a Möbius band or cross-cap, is correct.

c) a and b are to be identified with a^{-1} and b^{-1} in the configuration

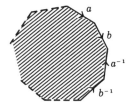

I claim that this is the result of sewing in a handle or tube, along two holes. Cut out the region in question, which is then

(where c is the line which we have cut). If we identify a and a^{-1}, we get a figure

Identifying b and b^{-1} (which effectively closes b^{-1} and glues it to b), we get a (punctured) torus glued to c.

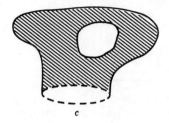

It's easy to check that the space obtained from identifying such a punctured torus to a hole in a sphere is homeomorphic to the space obtained from attaching a handle to two holes in a sphere, i.e.

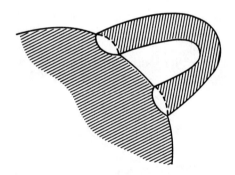

as claimed.

If we can show that M is equivalent to a disc, with these three simple kinds of identification, we shall be done. The rest of the proof is nothing but a systematic reduction to this form. We shall proceed by stages. Our disc D has edges which we write as, for example,

$$aba^{-1}b\,cc^{-1}.$$

We call this the *word* for the disc D. Each edge, with a given letter, occurs twice, because the only identifications glue two edges together. a means take a in a clockwise direction, a^{-1} means in a counterclockwise direction. Our steps reduce the word to a simpler form.

1) If there are at least four edges, any pair of edges aa^{-1}, that is a and a^{-1} are adjacent, may be eliminated from any word. This is best illustrated by a picture

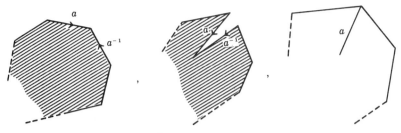

Note that we may write a^{-1} for a when we wish to indicate the reverse orientation, but for clarity both a^{-1} and the arrow are best.

In other words, if aa^{-1} occurs, we might just as well perform the identification. There remains a disc (the right figure) with two less edges.

2) I claim all vertices on the boundary of D, may be assumed to be identified together, when we form M from D.

If there are two vertices which are not identified, there must be two vertices which are adjacent and not identified, say A and B. We may assume our figure looks like

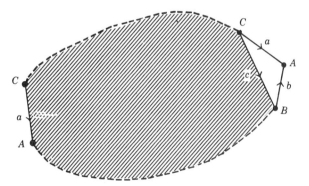

where we have added the line c from C to B. Note that our previous step allows us to assume that a, clockwise orientation, is not to be identified with b, counterclockwise orientation, so a must be identified with another edge, which we have drawn on the left.

We cut off the triangle ABC and glue it on the left

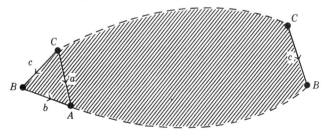

132 Topology

Since the two c's are to be identified, we have not changed the manifold or quotient space at all.

The new figure has one less vertex to be identified with A, one more identified with B. By repeating Step 1) above and then the above, over and over, we can reduce the number of vertices which is to be identified to a single given vertex A to none. One easily sees then that all vertices may be assumed to be identified. Naturally, the new disc, with identifications on the boundary, is quite different from the first, although the manifold M is the quotient space of each of them, up to a homeomorphism, of course.

3) Suppose a occurs twice, with the same exponent. For example

$$abcb^{-1}c^{-1}a$$

Then we may assume the two a's are adjacent. Consider the figure

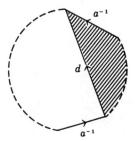

where we introduce d and the shading to explain the next step.

Cut along d, and glue the resulting figure along a^{-1}.

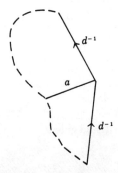

The result is a disc with suitable identifications still homeomorphic to M. No other sides, except a^{-1}, have been changed with respect to their order. But the two a^{-1}'s are replaced by two d^{-1} which are adjacent.

Clearly, we may repeat this till all such pairs are adjacent.

If all pairs are now adjacent, and of the same exponent, our earlier remarks (Example 8 and b) in this proof, above) show that M is a sphere with some holes removed and Möbius bands (or cross-caps) sewn in. This completes the proof in the case.

4) We now may assume that we have one pair of letters c and c^{-1} which are not adjacent. We claim that there is a d between c and c^{-1} with d^{-1} after c^{-1}; consider the figure

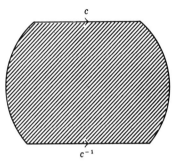

Since we have proved Step 3), all pairs a and a are adjacent. Suppose we have only adjacent pairs like aa in the right semicircle. Then no edge in the left semicircle can be identified with any edge in the right semicircle.

We immediately conclude that all vertices in the right semicircle are identified, and all the vertices in the left semicircle are identified. It then follows that the two vertices of c are not identified, contradicting Step 2).

Thus there is a d in the right semicircle, to be identified with d^{-1} in the left semicircle (it must be d^{-1}, for if it were d, d and d would be adjacent by Step 3).

5) We may transform D so that c and d, which occur as

$$\cdots c \cdots d \cdots c^{-1} \cdots d^{-1}$$

are together, i.e.

$$\cdots cdc^{-1}d^{-1} \cdots .$$

We illustrate this by a series of figures

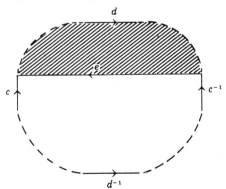

134   Topology

e and the shade region are introduced to explain the next figure,

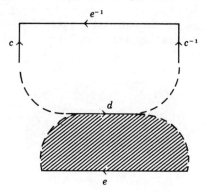

which we rewrite as

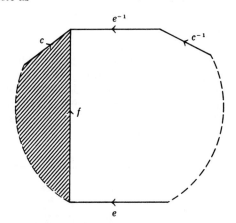

Once again, cut along f and glue along c.

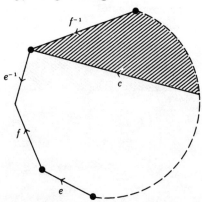

Manifolds and Classification of Surfaces **135**

This replaces $\cdots c \cdots d \cdots c^{-1} \cdots d^{-1}$ with $\cdots efe^{-1}f^{-1} \cdots$. One easily checks that those pairs $a \cdot a$ are not disturbed by this process.

We conclude that either Step 1) fails, in which case we have only two edges (and hence a sphere, as in a) of this proof, or a projective plane = sphere with a cross-cap, as in Example 7), or else we have a sphere with possible cross-caps (sewn in Möbius bands), as in b) in this proof, or handles, as in c) of this proof. Note that a cross-cap corresponds to aa and a handle to $cdc^{-1}d^{-1}$. This completes the proof.

Theorem 7.2 is not quite complete in that the description which it offers for compact, 2-manifolds, which are complexes, is not unique. The flexibility in the possible descriptions of the same 2-manifold is given as follows:

Proposition 7.6

Any three cross-caps (Möbius bands) attached to a 2-manifold may be removed, and one handle and one cross-cap added as replacements, with the resulting manifold remaining homeomorphic to the original manifold.

Proof. It will suffice to prove that:

a) a Möbius band with a handle attached (i.e. one handle and one cross-cap), and

b) a Möbius band with a Klein bottle attached (which, by Example 8, is equivalent to three Möbius bands to be attached to the manifold) are homeomorphic. (Check this out!)

We consider the following:

a)

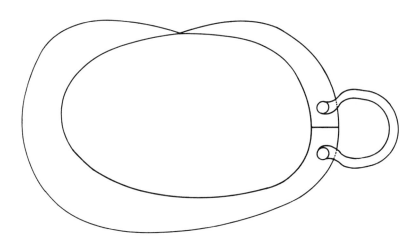

and b)

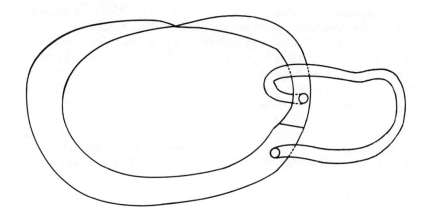

These figures represent a) and b) above. As the Möbius band has only one side, we see at once that they are homeomorphic, completing the proof. (Slide the handle, in case b), around the Möbius band a bit less than one whole revolution.)

Remarks. 1) We have one deficiency remaining in our treatment of 2-manifolds. That is, we have not proved that different 2-manifolds (for example the torus and the projective plane) are indeed *not* homeomorphic. This can be done best with the aid of the fundamental group (next chapter) and in fact, except for the equivalences arising from Proposition 7.6, all the manifolds discribed in Theorem 7.2 are distinct from one another, i.e. not homeomorphic.

2) The classification of higher dimensional manifolds poses serious problems. It has *not* been carried out even in dimension 3.

Problems

1. Identify the quotient of the disc, when the boundary is identified according to

    I) $abcc^{-1}b^{-1}a^{-1}$
    II) $acbcba$.

2. Assume the Jordan curve theorem (see texts of Dold, Spanier, etc.) that if $K \subseteq S^2$, with K homeomorphic to S^1, then

$$S^2 - K$$

   is disconnected.

   Prove that neither the torus nor the Klein bottle can be homeomorphic to S^2.

3. Show that there is a subset K of the projective plane P^2 which is homeomorphic to S^1, so that $P^2 - K$ is connected. Conclude with same assumptions as previous problem, that P^2 is not homeomorphic to S^2.
4. Show that every compact 2-manifold, which is a complex, is a differentiable manifold (Definition 7.5). (*Hint:* Prove that there is a cover of open sets, which overlap nicely, using the representation as a quotient of the disc, by identifications on the boundary, as in Theorem 7.2.)

CHAPTER 8

The Fundamental Group

We now wish to explore some further refinements which enable us to detect when certain spaces are not homeomorphic. Roughly speaking, if we wish to show that two spaces are homeomorphic, we have to construct a homeomorphism from one to the other. If we wish to show that two spaces are *not* homeomorphic, we must find an invariant—some construction which always gives the same result when applied to homeomorphic spaces—which is different on the two spaces in question. The fundamental group is one of the first, non-trivial, such constructions.

We wish to associate, with every space X, a group $\pi_1(X)$ so that whenever X and Y are homeomorphic, $\pi_1(X)$ and $\pi_1(Y)$ are isomorphic (for these concepts, see any basic text, such as Bourbaki *Algèbre*, etc.). Actually, we don't quite do this, for there are some technical details about base points which need consideration. Without further conditions, such a construction would do little to reflect the geometry involved, for the geometry or topology is as closely connected with the continuous mappings as with the spaces themselves. Thus, whenever

$$f: X \to Y$$

is a continuous map, we want to have a homomorphism

$$f_\#: \pi_1(X) \to \pi_1(Y).$$

Actually, a better name for $f_\#$ would be $\pi_1(f)$, but this form is generally accepted in the literature.

In fact, we shall construct a functor π_1 from a reasonable category of spaces to the category of groups (see "Remarks" at the end of Chapter 1). That means that if $1: X \to X$ is the identity, $1_\#$ is the identity homomorphism, and if we have continuous maps

$$X \xrightarrow{f} Y \xrightarrow{g} Z,$$

then $(g \cdot f)_\# = g_\# \cdot f_\#$, i.e. the operations of taking the associated homomorphism on fundamental groups and composing maps or homomorphisms are in fact interchangable.

The usefulness of this kind of construction cannot be too strongly emphasized, even though it does not appear to be very potent at first glance. In fact, we will see later that if we consider the circle and disc

$$S^1 = \{(x, y) \mid (x, y) \in \mathbb{R}^2, x^2 + y^2 = 1\}$$

and

$$D^2 = \{(x, y) \mid x^2 + y^2 \leq 1\}$$

(note $\bar{B}^2 = D^2$ in earlier notation), then

$$\pi_1(S^1) \approx Z, \text{ i.e. the group of integers}$$

and

$$\pi_1(D^2) \approx 0, \text{ the trivial group, with one element.}$$

(We write $\approx$ to mean that the groups are isomorphic.) Assuming these facts—to give a convincing application of these principles—we will sketch a proof of the famous Brouwer fixed-point theorem.

Brouwer Fixed-Point Theorem

Let $f: D^2 \to D^2$ be any continuous map of D^2 to itself.
Then there is a point

$$(x_0, y_0) \in D^2$$

such that

$$f((x_0, y_0)) = (x_0, y_0).$$

Proof. Assume this is not the case. For each point (x, y), draw the line from $f((x, y))$ to (x, y) until it meets the boundary S^1, calling the point where it meets the boundary $g((x, y))$. One easily checks
1) $g: D^2 \to S^1$ is continuous.
2) If $(x, y) \in S^1$, i.e. $x^2 + y^2 = 1$, then $g((x, y)) = (x, y)$.
Now, if $i: S^1 \to D^2$ is the inclusion map, i.e. $i((x, y)) = (x, y)$, we see that the composition

$$S^1 \xrightarrow{i} D^2 \xrightarrow{g} S^1$$

is the identity map $1: S^1 \to S^1$.

Apply π_1 to each term and each map. We see that

$$\pi_1(S^1) \xrightarrow{i_\#} \pi_1(D^2) \xrightarrow{g_\#} \pi_1(S^1)$$

must be the identity homomorphism of the integers to itself. But $\pi_1(D^2) = 0$, so $i_\#$ has an image consisting of one element, and thus $g_\# i_\#$ or $(g \cdot i)_\#$ has an image consisting of one element. Hence, the image of the identity map, from the integers to themselves consists of one element, which is absurd.

This contradiction proves the theorem.

There are many applications of this sort; but the fundamental group also plays a deep role in the theory of covering spaces (see the next chapter). As is frequently the case, there is some considerable amount of detail, which must be verified in the construction of such a group. Also common is the situation whereby the group is constructed as a quotient of something quite large (this is similar to singular homology, which is not treated in this book). We shall concentrate on the construction here; the calculations and hence applications are postponed until the last chapters.

Definition 8.1

Let X be an arcwise connected (Definition 4.6), topological space, and $x_0 \in X$ a fixed point, which we shall call the *base point*.

A loop based at x_0 is a continuous function

$$\alpha: I \to X$$

(recall $I = [0, 1]$) such that

$$\alpha(0) = \alpha(1) = x_0.$$

The set of all loops based at x_0 is denoted $\Omega(X, x_0)$ (and called the "set of loops based at x_0").

Clearly loops are basic geometric objects. The construction $\Omega(X, x_0)$ associates with X what frequently turns out to be a very big set $\Omega(X, x_0)$. But, we should note that this is a functor. For let $\mathcal{C}_*$ be the category whose objects are arcwise connected spaces with a base point chosen in each, and whose mappings are those continuous maps which send the base point of the domain to the base point of the range. Our functor Ω goes from such spaces to sets. If X is such a space, with base point x_0, our functor associates to X the set $\Omega(X, x_0)$. If $f: X \to Y$ is such a map (sends base point to base point), we define

$$\Omega(f): \Omega(X, x_0) \to \Omega(Y, y_0)$$

by

$$\Omega(f)(\alpha) = f \cdot \alpha$$

(check that this works). We thus have a functor from these particular spaces to sets, in the sense described at the end of Chapter 1. One easily checks that if $g: Y \to Z$ in $\mathcal{C}_*$, then

$$\Omega(g \cdot f) = \Omega(g) \cdot \Omega(f)$$

as well as $\Omega(1) = 1$, where this latter 1 means the identity map on sets.

The fundamental group will be constructed as a quotient of the set $\Omega(X, x_0)$. Whereas $\Omega(X, x_0)$ is just a set, on which we shall construct a multiplication, which is not a group, the quotient of $\Omega(X, x_0)$, by a suitable equivalence relation, will actually be a group. Here are the rough ideas.

A) If we are given two paths, both beginning and ending at x_0, we can run through both in order, getting a map from [0, 2] to X, which begins and ends at x_0. Cutting the domain down to half-size, i.e. reparametrizing to get a map from [0, 1] to X, we build a way of "multiplying" two loops and getting a third loop.

B) Two paths will be regarded as equivalent, if they may be deformed into one another (precise definition below), by a deformation which leaves the end points fixed at x_0. The set of equivalence, classes of loops will be the fundamental group, with the group operation coming from the multiplication defined on loops in A). In fact, we'll show that while $\Omega(X, x_0)$ only has a multiplication, but in general no inverses, the quotient by our equivalence relation actually has inverses and is a group.

Definition 8.2

Let $\alpha, \beta \in \Omega(X, x_0)$. Define $\alpha * \beta \in \Omega(X, x_0)$ by

$$(\alpha * \beta)(t) = \begin{cases} \alpha(2t), & \text{if } 0 \leq t \leq \frac{1}{2} \\ \beta(2t - 1), & \text{if } \frac{1}{2} \leq t \leq 1 \end{cases}$$

(check that this is well defined). This is the multiplication of loops, from A) above.

Definition 8.3

Let $\alpha, \beta \in \Omega(X, x_0)$. Define $\alpha \sim \beta$, if there is a continuous

$$F: I \times I \to X$$

such that
a) $F(t, 0) = \alpha(t)$
b) $F(t, 1) = \beta(t)$
c) $F(0, u) = F(1, u) = x_0$, all $u \in I$.

If $\alpha \sim \beta$, we say that f *is homotopic to* g *relative to* x_0. (Homotopic is a technical term meaning deformable to one another.) If $\alpha \sim \beta$, α can be slid into β, in a continuous way which leaves end points fixed.

The reader should check that this is an actual equivalence relation. For example, here is one step:

Suppose $\alpha \sim \beta$, $\beta \sim \gamma$, where this second equivalence relation is given by $G: I \times I \to X$.

Define $H: I \times I \to X$ by

$$H(t, u) = \begin{cases} F(t, 2u), & \text{if } 0 \leq u \leq \tfrac{1}{2} \\ G(t, 2u - 1), & \text{if } \tfrac{1}{2} \leq u \leq 1. \end{cases}$$

Because $F(t, 1) = \beta(t) = G(t, 0)$, H is a well defined continuous function, which shows $\alpha \sim \gamma$. This is the formal definition, alluded to in B) above.

Remark. More generally, if f and g map X to Y, $f(x_0) = g(x_0) = y_0$, we say f homotopic to g relative to x_0, or keeping x_0 fixed, if there is $F: X \times I \to Y$ with $F(x, 0) = f(x)$, $F(x, 1) = g(x)$, $F(x_0, u) = y_0$, all $u \in I$, $x \in X$.

We are now ready to build the fundamental group.

Theorem 8.1

Let $\Omega(X, x_0)$ be the set with multiplication $*$, and equivalence relation $\sim$ as defined above.

1) $*$ defines a multiplication on the quotient set

$$\Omega(X, x_0)/\sim$$

by $\{\alpha\} \cdot \{\beta\} = \{\alpha*\beta\}$, where the brackets mean the equivalence class of those elements, equivalent to the one inside the brackets.

2) With this multiplication from 1), the quotient set is a group (associative law, identity, and inverses).

3) If $f: X \to Y$, $f(x_0) = y_0$, the base point for Y, then

$$f_{\#}(\{\alpha\}) = \{f \cdot \alpha\}$$

defines a homomorphism from the group associated in 2) above with X, to that associated with Y.

We shall denote

$$\pi_1(X, x_0) = \Omega(X, x_0)/\sim$$

and call it the *fundamental group of X with base point x_0*. It is a functor on the category of arcwise connected spaces, with base points, and base point preserving maps to the category of groups.

Proof. Recall that the quotient set means the set of equivalence classes of elements of $\Omega(X, x_0)$, under this relation $\sim$. That is an element of the quotient set is the set of all elements in $\Omega(X, x_0)$ equivalent to a given element say α. We write $\{\alpha\}$ for this class of equivalent elements.

Recall also that $\{\alpha_1\} = \{\alpha_2\}$ precisely when $\alpha_1 \sim \alpha_2$.

We must show that when $\alpha_1 \sim \alpha_2$ and $\beta_1 \sim \beta_2$, then

$$\alpha_1*\beta_1 \sim \alpha_2*\beta_2.$$

For then the formula $\{\alpha\}\cdot\{\beta\} = \{\alpha*\beta\}$ is well defined on the equivalence classes (rather than just on the individual elements).

Recall that
$$\alpha_1*\beta_1(t) = \begin{cases} \alpha_1(2t), & 0 \leq t \leq \frac{1}{2} \\ \beta_1(2t-1), & \frac{1}{2} \leq t \leq 1 \end{cases}$$

and similarly for α_2 and β_2. Denote our two homotopies, $\alpha_1 \sim \alpha_2$ and $\beta_1 \sim \beta_2$ by
$$F: I \times I \to X \quad \text{and} \quad G: I \times I \to X.$$

We define a homotopy $\alpha_1*\beta_1 \sim \alpha_2*\beta_2$ as follows:
$$H(t, u) = \begin{cases} F(2t, u), & 0 \leq t \leq \frac{1}{2} \\ G(2t-1, u), & \frac{1}{2} \leq t \leq 1. \end{cases}$$

Notice that
$$H(t, 0) = \begin{cases} F(2t, 0), & 0 \leq t \leq \frac{1}{2} \\ G(2t-1, 0), & \frac{1}{2} \leq t \leq 1 \end{cases}$$
$$= \begin{cases} \alpha_1(2t); & 0 \leq t \leq \frac{1}{2} \\ \beta_1(2t-1); & \frac{1}{2} \leq t \leq 1 \end{cases}$$
$$= \alpha_1*\beta_1(t).$$

A similar computation will show at once that
$$H(t, 1) = \alpha_2*\beta_2(t).$$

(Check!) Furthermore,
$$H(0, u) = F(0, u) = x_0$$
and
$$H(1, u) = G(1, u) = x_0,$$

Thus, H is a homotopy $\alpha_1*\beta_1 \sim \alpha_2*\beta_2$, as I had claimed. Thus, the multiplication of classes in $\pi_1(X, x_0)$ is well-defined.

We must now carry out the work, admittedly rather tedious, to show that $\pi_1(X, x_0)$ is actually a group.

Let
$$\bar{e}: I \to X$$

be defined by $\bar{e}(t) = x_0$. In other terms, $\bar{e}$ is the constant loop. Set $e = \{\bar{e}\}$. I claim that e is the identity element of this group. To show this, consider the product
$$\{\alpha\}\cdot e: I \to X$$

which is defined as the class of $\alpha*\bar{e}$. Intuitively, $\alpha*\bar{e}$ is a loop which runs through α, at double speed, during the first half of I, then stays fixed at x_0. We visualize $I \times I$ as follows:

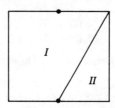

On the top, we wish to use the map α, on the bottom $\alpha*\bar{e}$. We will send all of region II to the base point x_0, and map region I to achieve our purpose. Define

$$F(t, u) = \begin{cases} \alpha\left(\dfrac{2t}{u+1}\right), & \text{if } u \geq 2t - 1 \\ x_0, & \text{if } u \leq 2t - 1 \end{cases} ;$$

note that $u = 2t - 1$ is the line dividing the regions I and II. If $u = 2t - 1$, $\alpha(2t/(u+1)) = \alpha(1) = x_0$, so that $F(t, u)$ is a continuous function. If $u \geq 2t - 1$,

$$0 \leq \frac{2t}{u+1} \leq 1$$

so that F is well-defined. But

$$F(t, 0) = \begin{cases} \alpha(2t), & \text{if } 2t \leq 1, \text{ i.e. } t \leq \tfrac{1}{2} \\ x_0 & \text{otherwise.} \end{cases}$$

Hence, by our definition,

$$F(t, 0) = \alpha*\bar{e}.$$

On the other hand,

$$F(t, 1) = \alpha(t).$$

Thus $\alpha \simeq \alpha*\bar{e}$, as desired. The proof that $e*\{\alpha\} = \{\alpha\}$ is nearly identical, using the figure

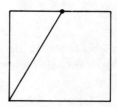

and should be checked out by the reader. A map which is homotopic to $\bar{e}$, leaving end points fixed, will be called *null homotopic*.

Next, we wish to show that $\pi_1(X, x_0)$ has inverses. Define

$$\bar{\alpha}(t) = \alpha(1 - t)$$

and

$$\{\alpha\}^{-1} = \{\bar{\alpha}\}.$$

We must verify $\{\alpha\} \cdot \{\alpha\}^{-1} = \{\alpha\}^{-1} \cdot \{\alpha\} = e$.

This is intuitively the following: $\alpha * \bar{\alpha}$ consists in running α and then α again in the reverse direction. The value of $\alpha * \bar{\alpha}(\frac{1}{2})$ is the base point, but since it is not the value at 0 or 1, it is free to move. We deform $\alpha * \bar{\alpha}$ to itself, by pulling it back through itself, i.e.

To make this formal, define a homotopy

$$F(t, u) = \begin{cases} \alpha(2t), & \text{if } t \leq \frac{1}{2}, \ u \leq 1 - 2t \\ \alpha(1 - u), & \text{if } t \leq \frac{1}{2}, \ u \geq 1 - 2t \\ \alpha(1 - u) = \bar{\alpha}(u), & \text{if } t \geq \frac{1}{2}, \ u \geq 2t - 1 \\ \bar{\alpha}(2t - 1), & \text{if } t \geq \frac{1}{2}, \ u \leq 2t - 1. \end{cases}$$

All the cases are well-defined, but we must check agreement on the overlapping regions. Suppose $t \leq \frac{1}{2}$ and $u = 1 - 2t$. Then in the first case, our definition is $\alpha(2t)$ and in the second case, it is $\alpha(1 - u) = \alpha(1 - (1 - 2t)) = \alpha(2t)$. Similarly, suppose $t = \frac{1}{2}$; then, clearly, the second and third cases are identical. Lastly, suppose $t \geq \frac{1}{2}$, $u = 2t - 1$. Then $\bar{\alpha}(u) = \bar{\alpha}(2t - 1)$, completing the verification that $F(t, u)$ is a well-defined and continuous function.

Notice that

$$F(t, 0) = \alpha * \bar{\alpha}$$

while

$$F(t, 1) = \begin{cases} \alpha(0), & \text{if } 1 \leq 1 - 2t, \text{ or } t = 0 \\ \alpha(0), & 0 < t < \frac{1}{2} \\ \bar{\alpha}(1), & \frac{1}{2} \leq t < 1 \\ \bar{\alpha}(1), & t = 1. \end{cases}$$

Hence, $\alpha * \bar{\alpha} \simeq \bar{e}$. The same sort of argument yields $\bar{\alpha} * \alpha \simeq \bar{e}$.

Topology

Lastly, we must show associativity, that is $(\{\alpha\} \cdot \{\beta\}) \cdot \{\gamma\} = \{\alpha\} \cdot (\{\beta\} \cdot \{\gamma\})$, or in other words

$$(\alpha * \beta) * \gamma \simeq \alpha * (\beta * \gamma).$$

This is the most complex of these verifications. Using the picture

define

$$F(t,u) = \begin{cases} \alpha\left(\dfrac{4t}{1+u}\right), & \text{if } 0 \leq t \leq \dfrac{u+1}{4} \\ \beta(4t - 1 - u), & u+1 \leq 4t \leq u+2 \\ \gamma\left(1 - \dfrac{4(1-t)}{2-u}\right), & \dfrac{u+2}{4} \leq t \leq 1. \end{cases}$$

One easily checks that each piece is well-defined, and we must consider the overlapping regions. If $4t = u + 1$

$$\alpha\left(\frac{4t}{1+u}\right) = \alpha(1) = x_0$$

$$\beta(4t - 1 - u) = \beta(0) = x_0.$$

If $4t = u + 2$,

$$\beta(4t - 1 - u) = \beta(1) = x_0,$$

while

$$\gamma\left(1 - \frac{4(1-t)}{2-u}\right) = \gamma\left(1 - \frac{4-u-2}{2-u}\right)$$

$$= \gamma\left(\frac{2-u-4+u+2}{2-u}\right) = \gamma(0) = x_0.$$

But then

$$F(t,0) = \begin{cases} \alpha(4t), & \text{if } 0 \leq t \leq \tfrac{1}{4} \\ \beta(4t - 1), & \tfrac{1}{4} \leq t \leq \tfrac{1}{2} \\ \gamma(2t - 1), & \tfrac{1}{2} \leq t \leq 1 \end{cases}$$

whereas
$$F(t, 1) = \begin{cases} \alpha(2t), & \text{if } 0 \le t \le \frac{1}{2} \\ \beta(4t-2), & \text{if } \frac{1}{2} \le t \le \frac{3}{4} \\ \gamma(4t-3), & \text{if } \frac{3}{4} \le t \le 1. \end{cases}$$

It is absolutely routine to verify that these are $(\alpha*\beta)*\gamma$ and $\alpha*(\beta*\gamma)$, respectively, showing the associative law. (Please check details here.)

Finally, we need check that

$$f_\#: \pi_1(X, x_0) \to \pi_1(Y, y_0)$$

defined by $f_\#\{\alpha\} = \{f \cdot \alpha\}$ is a homomorphism. (Note that if $\alpha_1 \simeq \alpha_2$, by a homotopy F, then $f\alpha_1 \simeq f\alpha_2$ by homotopy $f \cdot F$, showing that $f_\#$ is well-defined). We must therefore prove that

$$f_\#(\{\alpha\} \cdot \{\beta\}) = (f_\#(\{\alpha\})) \cdot (f_\#(\{\beta\}))$$

or

$$f \cdot (\alpha*\beta) \simeq (f \cdot \alpha)*(f \cdot \beta).$$

In fact, I claim that $f \cdot (\alpha*\beta) = (f\alpha)*(f\beta)$. For

$$f \cdot (\alpha*\beta) = \begin{cases} f(\alpha(2t)), & \text{if } 0 \le t \le \frac{1}{2} \\ f(\beta(2t-1)), & \text{if } \frac{1}{2} \le t \le 1 \end{cases}$$

while $(f \cdot \alpha)*(f \cdot \beta)$ is trivially checked to be that same map. Hence, $f_\#$ is a homomorphism.

Remarks. 1) With no more difficulty, we may check that whenever $f: X \to Y$, $g: Y \to Z$, $f(x_0) = y_0$, $g(y_0) = z_0$, then $(g \cdot f)_\#$ and $g_\# \cdot f_\#$ are the same homomorphism from $\pi_1(X, x_0)$ to $\pi_1(Z, z_0)$. Clearly, $1_\#$ is the identity homomorphism from $\pi_1(X, x_0)$ to itself. π_1 is thus a functor, as defined in Chapter 1.

2) We do not yet have the tools to calculate $\pi_1(X, x_0)$ in interesting cases. When we do achieve these methods, in the next chapters, we shall see that in general $\{f\} \cdot \{g\} \ne \{g\} \cdot \{f\}$, i.e. $\pi_1(X, x_0)$ need not be an Abelian group. We will, however, show that this is the case, when X is a topological group. (Compare Definition 7.3.)

We begin now with some elementary remarks.

Proposition 8.1

Suppose X and Y are arcwise connected spaces, with base points x_0 and y_0. Suppose $f: X \to Y$ is a homeomorphism such that $f(x_0) = y_0$. Then

$$f_\#: \pi_1(X, x_0) \to \pi_1(Y, y_0)$$

is an isomorphism (a 1 to 1 and onto homomorphism).

Proof. $f^{-1}: Y \to X$, defined by $f^{-1}(y)$ = the unique x such that $f(x) = y$, is a homeomorphism with $f^{-1}(y_0) = x_0$. Also $f^{-1} \cdot f = 1_X$ and $f \cdot f^{-1} = 1_Y$.

It follows at once (from Remark 1 above) that

$$f_\#: \pi_1(X, x_0) \to \pi_1(Y; y_0)$$

and

$$(f^{-1})_\#: \pi_1(Y, y_0) \to \pi_1(X, x_0)$$

are inverse homomorphisms (in that their compositions are equal to the identity). Hence, each is 1-1 and onto, and in particular $f_\#$ is an isomorphism.

Proposition 8.2

Let X be an arcwise connected space, x_0, x_1 two points of X. Let ρ be a path (i.e. a map $I \to X$), beginning and ending at x_0 and x_1, respectively. Then ρ defines an isomorphism

$$\phi_\rho: \pi_1(X; x_0) \to \pi_1(X; x_1).$$

Proof. Define, for a path $\alpha \in \Omega(X; x_0)$

$$\bar{\phi}_\rho(\alpha)(t) = \begin{cases} \rho(1 - 3t), & 0 \le t \le \tfrac{1}{3} \\ \alpha(3t - 1), & \text{if } \tfrac{1}{3} \le t \le \tfrac{2}{3} \\ \rho(3t - 2), & \text{if } \tfrac{2}{3} \le t \le 1. \end{cases}$$

We might picture this as

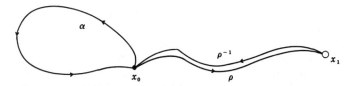

One easily checks that $\alpha_1 \simeq \alpha_2$ implies that $\bar{\phi}_\rho(\alpha_1) \simeq \bar{\phi}_\rho(\alpha_2)$. Set

$$\phi_\rho(\{\alpha\}) = \{\bar{\phi}_\rho(\alpha)\}$$

(Check that this is well-defined.)

Note that $\alpha*\beta$, with $\alpha \in \Omega(X, x_0)$, $\beta \in \Omega(X, x_0)$ may be deformed (with end points left fixed) to a path which runs through α, then ρ, then ρ^{-1} (given by $\rho^{-1}(t) = \rho(1 - t)$) and lastly β. This is illustrated by the following pictures

(Check the details here; it is just like most of the previous constructions.)
One then shows, without difficulty (but with considerable detail!) that

$$\phi_\rho(\{\alpha\}\cdot\{\beta\}) = \phi_\rho(\{\alpha\})\cdot\phi_\rho(\{\beta\}).$$

All this may be repeated for ρ^{-1}, $(\rho^{-1}(t) = \rho(1-t))$. Consider

$$\phi_{\rho^{-1}}\cdot\phi_\rho(\{\alpha\}) = \phi_{\rho^{-1}}\{\bar\phi_\rho(\alpha)\}.$$

One may work this out, for each time t, by our definition of ρ. But it is much more conceptual to notice that this consists of the homotopy class of the path which

  i) begins at x_0, and runs to x_1, along ρ.
  ii) runs back along ρ^{-1} to x_0
  iii) passes around α
  iv) returns to x_1 along ρ
  v) returns to x_0 along ρ^{-1}.

Clearly, paths which run ρ followed by ρ^{-1} may be shrunk to a point (i.e. initial point of ρ). One then checks that $\phi_{\rho^{-1}}\cdot\phi_\rho(\{\alpha\}) = \{\alpha\}$ (without difficulty).

Hence, we see that ϕ_ρ has an inverse (in fact, an inverse on either side). Thus ϕ_ρ is an isomorphism.

I have deliberately left out some details here because they are so similar to previous constructions and I want the reader to figure some of these out from first principles. Note that the homomorphism ϕ_ρ depends on the path ρ itself, not just the end points x_0 and x_1.

Corollary 8.1

If X and Y are homeomorphic spaces which are arcwise connected, then $\pi_1(X, x_0)$ is isomorphic to $\pi_1(Y, y_0)$, for any $x_0 \in X, y_0 \in Y$.

(Check that this follows from the preceding two propositions!)

Definition 8.4

Let X be an arcwise connected space, $x_0 \in X$. Suppose that the identity map $1: X \to X$ is homotopic to the map $p: X \to X$ defined by $p(x) = x_0$, for all $x \in X$. We assume, as in Definition 8.3(c), that for this homotopy, $F(x_0, u) = x_0$, all $u \in I$ (here $F: X \times I \to X$).

Then we say that X is *contractible to* x_0.

Examples. 1) A single point space x_0 is trivially contactible to x_0.

2) Let $x_0 \in \mathbb{R}$ be any real number. Then $\mathbb{R}$ is contractible to x_0. Define $F(x, u) = u(x - x_0) + x_0$. Then, $F(x, 0) = x_0$ for all x; $F(x, 1) = x$, and for any $u \in I, f(x_0, u) = x_0$.

3) If X is contractible to x_0 and Y is contractible to y_0, then $X \times Y$ is contractible to (x_0, y_0).

Corollary 8.2

If X is contractible to x_0, $\pi_1(X, x_0)$ is the trivial group (the sole element being the identity, e).

Proof. Let $i: x_0 \to X$ be the inclusion. Then, by Definition 8.4,

$$x_0 \xrightarrow{i} X \xrightarrow{\rho} x_0$$

is the identity map. On the other hand,

$$X \xrightarrow{\rho} x_0 \xrightarrow{i} X$$

is homotopic to the identity map, keeping the base point fixed (this is the definition of contractible).

We see, from our main Theorem 8.1, that

$$i_\#: \pi_1(x_0, x_0) \to \pi_1(X, x_0)$$

and

$$\rho_\#: \pi_1(X, x_0) \to \pi_1(x_0, x_0)$$

are homomorphisms and $\rho_\# \cdot i_\#$ is the identity. We shall prove (our next proposition) that if a map f is homotopic to another g, keeping the base point fixed, then they both define the same map (or induce the same map) on the fundamental group, that is $f_\# = g_\#$. Therefore, $i_\# \cdot \rho_\#$ is also the identity.

It follows that $i_\#$ and $\rho_\#$ are inverse maps, each being an isomorphism. Then

$$\pi_1(X, x_0) \approx \pi_1(x_0, x_0).$$

But $\pi_1(x_0, x_0)$ is the trivial group (check!). (We occasionally write $\pi_1(x_0, x_0) = \{e\}$, where $\{e\}$ means the set consisting of only the identity element.)

Proposition 8.3

Let f and g map X to Y, $f(x_0) = g(x_0) = y_0$. Suppose f and g are homotopic, relative to x_0. That is, we have $F: X \times I \to Y$, with $F(x, 0) = f(x)$, $F(x, 1) = g(x)$, and $F(x_0, u) = y_0$, all $u \in I$, $x \in X$.

Then $f_\#$ and $g_\#$ are the exact same homomorphism from $\pi_1(X, x_0)$ to $\pi_1(Y, y_0)$. That is

$$f_\#(\{\alpha\}) = g_\#(\{\alpha\}), \quad \text{all} \quad \{\alpha\} \in \pi_1(X, x_0).$$

Proof.

$$f_\#(\{\alpha\}) = \{f \cdot \alpha\}$$
$$g_\#(\{\alpha\}) = \{g \cdot \alpha\}.$$

Since f and g are homotopic, keeping x_0 fixed, by a homotopy F, and $\alpha: I \to X$, $\alpha(0) = \alpha(1) = x_0$, the map

$$F(\alpha(t), u)$$

is clearly a homotopy of $f \cdot \alpha$ and $g \cdot \alpha$, keeping 0 and 1, in the domain space I, fixed at y_0 in the range. Thus $\{f \cdot \alpha\} = \{g \cdot \alpha\}$.

Remarks. 1) $\pi_1(R^n; x_0) = \{e\}$, for any $n \geq 0$. For $n = 0$ this is trivial. For $n \geq 1$, this follows from the fact that R is contractible, and the product of contractible spaces is contractible.

This suggests, perhaps vaguely, that there should be a simple way to understand the fundamental group of a product (see Proposition 8.4 below).

2) When we establish (in the final chapter) various non-trivial fundamental groups for a variety of interesting spaces, we shall have an effective way of determining that maps are not homotopic, by showing that they yield distinct homomorphisms of the fundamental group.

Proposition 8.4

Assume X and Y are arcwise connected, with base points x_0 and y_0. Then

$$\pi_1(X \times Y; x_0 \times y_0) \approx \pi_1(X, x_0) \times \pi_1(Y, y_0)$$

(the term on the right means the direct product of the two groups, or the group whose elements are ordered pairs of elements, the first from $\pi_1(X, x_0)$, the second from $\pi_1(Y; y_0)$).

Proof. Define

$$\phi: \pi_1(X \times Y; x_0 \times y_0) \to \pi_1(X; x_0) \times \pi_1(Y, y_0)$$

by $\phi(\{\alpha\}) = (\{\pi_1 \cdot \alpha\}, \{\pi_2 \cdot \alpha\})$, where $\pi_1: X \times Y \to X$ and $\pi_2: X \times Y \to Y$ are the projections (take care not to confuse the two uses of π_1). (Check that this is a well-defined homomorphism!)

Define

$$\psi: \pi_1(X, x_0) \times \pi_1(Y, y_0) \to \pi_1(X \times Y, x_0 \times y_0)$$

by $\psi(\{\beta\}, \{\gamma\}) = \{\delta\}$, where δ is defined by

$$\delta(t) = (\beta(t), \gamma(t)).$$

(Check!)

We calculate

$$(\psi \cdot \phi)\{\alpha\} = \psi(\phi\{\alpha\})) = \psi(\{\pi_1\alpha\}, \{\pi_2\alpha\}) = \{(\pi_1\alpha, \pi_2\alpha)\}.$$

But $\alpha: I \to X \times Y$ is precisely the map $t \to (\pi_1 \cdot \alpha(t), \pi_2 \cdot \alpha(t))$. (Check! Compare introduction concerning products.) Hence, $\psi \cdot \phi =$ the identity.

On the other hand,

$$(\phi \cdot \psi)(\{\beta\}, \{\gamma\}) = \phi(\psi(\{\beta\}, \{\gamma\})) = \phi(\{\beta, \gamma\}) = (\{\beta\}, \{\gamma\}),$$

because $\pi_1(\beta(t), \gamma(t)) = \beta(t)$, while $\pi_2(\beta(t), \gamma(t)) = \gamma(t)$.

Thus ϕ and ψ are inverse isomorphisms, and the result follows.

Remark. When we establish $\pi_1(S^1, x_0) \approx Z$, the group of integers, we will conclude immediately that $\pi_1(\text{torus}) = \pi_1(S^1 \times S^1, x_0 \times x_0) \approx Z \times Z$, etc.

We are now ready for a more specific result.

Theorem 8.2

Let X be a topological group, arcwise connected with identity element e. Then $\pi_1(X, e)$ is commutative, or Abelian (that is $\{\alpha\} \cdot \{\beta\} = \{\beta\} \cdot \{\alpha\}$, for all $\{\alpha\}, \{\beta\} \in \pi_1(X, e)$). (We are using e for the base point.)

Proof. Recall that if X is a topological group, we have a map

$$\mu: X \times X \to X$$

so that $\mu(x, e) = \mu(e, x) = x$, for every $x \in X$.

Consider the unit square I^2 in the plane, i.e.

$$I^2 = \{(x, y) \mid 0 \leq x, y \leq 1\}.$$

The boundary of I^2 is made up of four intervals

$$\begin{aligned} A &= \{(x, 0) \mid 0 \leq x \leq 1\} \\ B &= \{(1, y) \mid 0 \leq y \leq 1\} \\ C &= \{(x, 1) \mid 0 \leq x \leq 1\} \\ D &= \{(0, y) \mid 0 \leq y \leq 1\}. \end{aligned}$$

Draw a picture! We use ∂I^2 to denote this boundary.

If $\{\alpha\}, \{\beta\} \in \pi_1(X, e)$, we shall form a map of

$$\partial I^2 = A \cup B \cup C \cup D$$

to X which maps the four intervals according to $\{\alpha\}, \{\beta\}, \{\alpha\}^{-1}$, and $\{\beta\}^{-1}$. Specifically, we define

$$\phi: \partial I^2 \to X$$

by

$$\phi(x, y) = \begin{cases} \alpha(x), & \text{if } y = 0, 1 \\ \beta(y), & \text{if } x = 0, 1. \end{cases}$$

(Check that this is well-defined.) Define a map

$$\rho: I \to \partial I^2$$

by
$$\rho(t) = \begin{cases} (2t, 0), & 0 \leq t \leq \tfrac{1}{2} \\ (1, 4t - 2), & \tfrac{1}{2} \leq t \leq \tfrac{3}{4} \\ (7 - 8t, 1), & \tfrac{3}{4} \leq t \leq \tfrac{7}{8} \\ (0, 8 - 8t), & \tfrac{7}{8} \leq t \leq 1. \end{cases}$$

Then the map $\phi \cdot \rho \colon I \to X$ represents

$$\{\alpha\} \cdot (\{\beta\} \cdot (\{\alpha\}^{-1} \cdot \{\beta\}^{-1})).$$

(Check this!) But $\pi_1(X, e)$ is associative, so we easily see that $\phi \cdot \rho$ is homotopic to a map which devotes a quarter of the interval to α, β, $\bar{\alpha}$, and $\bar{\beta}$, which we also denote $\phi \cdot \rho$. It will suffice, in order to prove $\pi_1(X, e)$ is commutative, to prove that $\phi \cdot \rho$ is homotopic to the constant map $\bar{e} \colon I \to X$, $\bar{e}(t) = e$. This is because $\{\alpha\} \cdot \{\beta\} \cdot \{\alpha\}^{-1} \cdot \{\beta\}^{-1} = e$ implies $\{\alpha\} \cdot \{\beta\} = \{\beta\} \cdot \{\alpha\}$. To this end, we define

$$\Phi \colon I^2 \to X$$

by

$$\Phi(x, y) = \mu(\alpha(x), \beta(y))$$

so that $\Phi(x, y) = \phi(x, y)$, whenever $(x, y) \in \partial I^2$. If $(x, y) \in I^2$, we write it in polar coordinates $\{r(x, y), \theta(x, y)\}$. Then we put

$$F(t, u) = \Phi(u \cdot (r(\rho(t)), \theta(\rho(t))).$$

Continuity is no problem, but we need to calculate

$$F(t, 0) = \Phi \text{ (the origin)} = \Phi(0, 0) = \mu(\alpha(0) \cdot \beta(0)) = \mu(e, e) = e.$$

$$F(t, 1) = \Phi(\{r(\rho(t), \theta(\rho(t))\}).$$

But $\rho(t) \in \partial I^2$; hence, we have

$$F(t, 1) = \phi(\rho(t)).$$

That is, $\phi(\rho(t)) \simeq \alpha * (\beta * (\bar{\alpha} * \bar{\beta}))$, or

$$F(t, 1) \in \{\alpha\} * (\{\beta\} * (\{\alpha\}^{-1} * \{\beta\}^{-1})).$$

Finally,

$$F(0, u) = \Phi(u \cdot (r(\rho(0)), \theta(\rho(0)))) = \Phi(u \cdot 0, 0) = e$$

and similarly

$$F(1, u) = e.$$

This shows $\phi \cdot \rho \simeq \bar{e}$, completing the proof.

Remarks. 1) We have not fully used the hypothesis that X is a topologi-

154   Topology

cal group; for example, we never used the associativity of X or the existence of inverses. If a space satisfies the axioms, for a topological group, which are used here, it is called an H space (after H. Hopf, who initiated the study of this important class of spaces).

2) The key to the proof is the fact that the map $\phi: \partial I^2 \to X$ may be extended to $\Phi: I^2 \to X$. This makes strong use of the fact that X has a multiplication μ. It is not possible in general.

3) If in fact $\pi_1(X, x_0)$ is not Abelian, then we conclude that X cannot be a topological group (or even an H space). We shall see, for example, in the final chapter, that the "figure-eight," i.e. the following subspace of the plane

$$\{(x, y) \mid (x - 1)^2 + y^2 = 1 \quad \text{or} \quad (x + 1)^2 + y^2 = 1\}$$

has non-Abelian fundamental group, and thus is not a topological group.

Problems

1. Prove that under the assumption that there is some space with non-trivial fundamental group, then it follows that

$$\pi_1(S^1; x_0)$$

is non-trivial (for any fixed $x_0 \in S^1$, the circle). (*Hint:* Any map $\alpha: I \to X$, with $\alpha(0) = \alpha(1) = x_0$, factors through S^1; i.e. there are $\alpha': I \to S^1$, and $\alpha'': S^1 \to X$ with $\alpha = \alpha'' \cdot \alpha'$).

2. Prove that under the assumption that there is some space, with non-commutative fundamental group, the "figure-eight" space has a non-commutative fundamental group. (*Hint:* A commutator of two elements, that is something of the form $\{\alpha\} \cdot \{\beta\} \cdot \{\alpha\}^{-1} \cdot \{\beta\}^{-1}$, will factor through the "figure-eight" space).

3. Let D^2 be the disc of unit radius in $\mathbb{R}^2$, $x_0 = (1, 0)$; let $S^1 = \partial D^2$ be the unit circle. Prove that a map $\phi: S^1 \to Y$, $\phi(x_0) = y_0$, is homotopic to the constant map $\bar{e}: S^1 \to Y$ defined by $\bar{e}(x) = y_0$, if and only if there is an extension $\Phi: D^2 \to Y$, that is a continuous map which, when restricted to S^1, agrees with ϕ.

4. Let $j: S^1 \to S^1$ be reflection in the x-axis; $x_0 = (1, 0)$. Then we have a homomorphism

$$j_\#: \pi_1(S^1, x_0) \to \pi_1(S^1; x_0).$$

Prove $\{\alpha\} + j_\#\{\alpha\} = 0$, all $\{\alpha\} \in \pi_1(S^1; x_0)$. Here, we write the group additively, and the identity element as 0, in anticipation of the fact that we shall soon show that $\pi_1(S^1; x_0)$ is isomorphic to the integers.

CHAPTER 9

Covering Spaces

We now study the theory of covering spaces. This is a classic and beautiful theory, whose origins come from analysis and predate much of topology. It studies spaces which—in a rough sense—are locally like another space, but have a continuous projection onto that other space. For example, if we imagine the real numbers as a helix in space, then there is a nice projection onto the circle

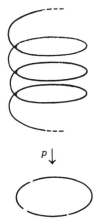

(p may be expressed precisely in terms of exponentials, as we do later) Locally, p appears to be a homeomorphism, because it sends small open intervals onto small open intervals of S^1, and the inverse image of a small open interval is a disjoint union of same. But, of course, the map p is far from being 1-1. This is a classic covering space.

It is obvious that one can study such spaces for their own geometric beauty. But then why look at them at this point of our book? The reason for their study here is a somewhat surprising fact: the classification of covering spaces of a given space X depends on the fundamental group $\pi_1(X; x_0)$. Roughly speaking, there are as many distinct covering spaces of X as there are distinct subgroups of $\pi_1(X)$. It shows the power of the fundamental group functor as a method of distinguishing topological things by their algebraic properties.

In another—perhaps surprising—development, one may use covering spaces to effectuate actual calculations of $\pi_1(X, x_0)$. We do a bit of this here, although the bulk of our calculations are postponed to the last chapter, where I describe, for suitable spaces, an algorithmic approach to the determination of $\pi_1(X, x_0)$.

The results of this chapter are all easy, and follow in a straightforward way from previous results, with one exception. That consists of our (last) result on the existence of covering spaces. To do this in its natural generality, we need a short point-set theoretic digression to look at "semi-locally simply connected" spaces. This is seemingly unreasonable, but this new concept turns out to be just the right one for our theory, and does not take an unreasonable amount of time to develop. We begin with definitions.

Definition 9.1

X is *locally arcwise connected*, if whenever

$$x \in U \subseteq X, \quad U \text{ open,}$$

then there is an arcwise connected, open V, with

$$x \in V \subseteq U \subseteq X.$$

For example, any open subset of R^n will be locally arcwise connected. More generally, any n-manifold will be locally arcwise connected, even though it may not be arcwise connected. On the other hand, the space

$$X = A \cup B \cup C \cup D$$

with

$$A = \{(x, 0) \mid -2 \leq x \leq 0\}$$
$$B = \{(0, y) \mid -1 \leq y \leq 1\}$$
$$C = \{(x, y) \mid y \leq 0, x^2 + y^2 = 4\}$$
$$D = \left\{(x, y) \mid y = \sin \frac{4\pi}{x}, 0 < x \leq 2\right\},$$

which is a closed up version of the graph of $\sin 1/x$, i.e.

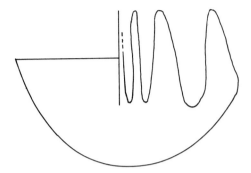

is arcwise connected (check!) but not locally arcwise connected (check the neighborhood at the origin). (Compare Definition 4.6.) Some authors call this a "topologist's sine curve."

In this chapter, we *assume* all our spaces to be both locally arcwise connected and arcwise connected Hausdorff spaces.

Definition 9.2

Let $p: \tilde{X} \to X$ be a continuous map of $\tilde{X}$ onto X, where both spaces are both locally arcwise connected and arcwise connected Hausdorff spaces.

Then we say the $\tilde{X}$ is a *covering space* of X, with projection map p, provided for each $x \in X$, there is a connected open set U, $x \in U$, so that

$$p^{-1}(U) = \bigcup_{\alpha \in A} V_\alpha, \quad V_\alpha \subseteq X \text{ being open,}$$

with
a) $V_\alpha \cap V_\beta = \phi$, if $\alpha \neq \beta$, $\alpha, \beta \in A$, A being simply the set of all such V_α's.
b) $p \mid V_\alpha : V_\alpha \to U$ is a homeomorphism for each $\alpha \in A$, where $p \mid V_\alpha$ means p restricted to the subset V_α of $\tilde{X}$, that is p applied only to elements of V_α. (Note that if there is such a U, then we can assume that we have such a set U, which is arcwise connected).

Remarks and Examples. 1) If $\tilde{X} \equiv X$, $p = $ the identity, and X is a locally arcwise connected and arcwise connected Hausdorff space, then $\tilde{X}$ is a trivial covering space of X. (Take $U = X$, and A to be set with one element.)

2) If X is a single point, then 1) above is the only possible covering space.

3) Let $X = S^1$, the circle. We then have many possible covering spaces. For example.
   a) Let $\tilde{X} = S^1$, $p_1: \tilde{X} \to X$ be defined by $p_1(e^{i\theta}) = e^{4i\theta}$, representing S^1 as the unit circle in the complex plane. It is easy to tabulate p_1 and you may quickly find the value of p_1 for several points in $\tilde{X}$.

158   Topology

We thus conclude that p wraps the circle around itself, in a linear way, 4 times. If we take a small open neighborhood U of any point $x \in S^1 = x$, then $p^{-1}(U)$ is a disjoint union of 4 copies of U (up to homeomorphism), as desired. We refer to this as a 4-sheeted cover, as p^{-1} (point) is 4 points.

b) Given any integer $n > 0$, the map $p(e^{i\theta}) = e^{ni\theta}$ defines an n-sheeted covering space of S^1, where p^{-1} (point) consists of n points.

c) The map $\pi \colon \mathbb{R} \to S^1$ defined by $\pi(x) = e^{2\pi i x}$ maps $\mathbb{R}$ onto S^1 in such a way that each interval $[n, n+1]$ is mapped onto S^1. For a small open interval $U \subseteq S^1$, that is any open interval which is not all of S^1, we have

$$p^{-1}(U) = \bigcup_{\alpha \in A} V_\alpha$$

where A is the integers and each V_α is homeomorphic to U. We illustrate this as follows:

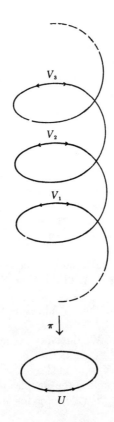

This all shows that there are many different possibilities for covering spaces, some of which may be compact (Cases a and b) and some of which need not be compact (Case c).

4) The sphere is a covering space of the projective plane (Definition 7.2).

5) The definition of a covering space is a strong, sort of uniformity on $\tilde{X}$. For example, in 3(c), above, replace $\mathbb{R}$ by $\langle 0, 2 \rangle$. Then

$$\pi: \langle 0, 2 \rangle \to S^1$$

is onto, but *not* a covering space. (Check $\pi^{-1}(U)$ when U is an open set containing $(1, 0) \in S^1$.)

6) There are a wealth of other examples from the theory of 2-manifolds, or surfaces. For example, the torus is a twofold covering of the Klein bottle. The plane is a covering space of the torus. In fact, the torus is a quotient of the topological group $\mathbb{R}^2 = \mathbb{R} \oplus \mathbb{R}$, that is the group of pairs of real numbers with pairwise addition, by the subgroup consisting of all pairs of integers. This is quickly checked to be a covering space (see "Problems" at the end of this chapter).

7) It shall follow from our later theorems that X (satisfying our usual conditions) admits *no* proper covering space, if and only if $\pi_1(X, x_0)$ is the trivial group (i.e. one element e).

8) In a covering space, $p^{-1}(x)$ is discrete for any $x \in X$. (Check!)

In order to begin our study of covering spaces, we need to know when a path in X yields another path in $\tilde{X}$ which lies over it. The following is a very general answer.

Proposition 9.1

Let $p: \tilde{X} \to X$ be a covering space. Let $\alpha: I \to X$ be a path in X. Suppose $\alpha(0) = x_0.\ p(\tilde{x}_0) = x_0$.
Then there is a unique path

$$\tilde{\alpha}: I \to \tilde{X}$$

such that $\tilde{\alpha}(0) = \tilde{x}_0$, and $p \cdot \tilde{\alpha}(t) = \alpha(t)$ for all $t \in I$.

The gist of this is that once you fix an initial point over x_0, i.e. $\tilde{x}_0$, the path lifts uniquely. (This is easily checked in the above examples!)

Proof. Choose a cover of $\alpha(I) \subseteq X$ by a finite number of open sets U_j, of the sort occurring in our definition of a covering space. Assume that the U_j's are ordered so that they cover $\alpha(I)$ in the order of increasing values of t., as much as this is possible.

For example,

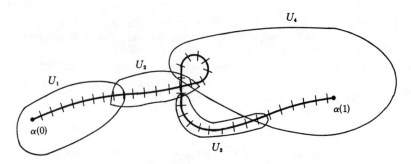

where we write the path as ++++++++. We would like to work with these U_j's, but unfortunately some technical difficulties might arise if a U_j meets the path $\alpha(I)$ in several distinct pieces. However, I claim that we may choose U_j's so that each $U_j \cap \alpha(I)$ is connected, in addition to the above assumptions on the U_j.

Every open set in I is a disjoint union of open intervals. I is compact, as is every closed subset. The image under α of a closed set is then closed, and its complement will be open. Each open U_i may then be replaced, if necessary, with a smaller open set which meets $\alpha(I)$ in the image of a small interval.

Note that the condition that $p^{-1}(U_j)$ is a disjoint union of homeomorphic copies of U_j is preserved on passing to smaller sets. We then check without difficulty that we may find t_i,

$$0 = t_0 \leq t_1 \leq \cdots \leq t_j \leq \cdots \leq t_n \leq t_{n+1} = 1,$$

so that each $\alpha([t_{m-1}, t_m])$ lies entirely within some U_j for which our condition holds. Each $\alpha([t_{m-1}, t_m])$ is connected, being the image of a connected set.

I will now show that given a value for $\tilde{\alpha}(t_{m-1})$, such that $p \cdot \tilde{\alpha}(t_{m-1}) = \alpha(t_{m-1})$ there is a unique continuous $\tilde{\alpha}(t)$, for $t_{m-1} \leq t \leq t_m$ so that

$$p \cdot \tilde{\alpha}(t) = \alpha(t), \quad \text{for} \quad t_{m-1} \leq t \leq t_m.$$

Because $\alpha(t)$, $t_{m-1} \leq t \leq t_m$ lies in U_m, and

$$p^{-1}(U_m) = \bigcup_\beta V_\beta,$$

we see that $\tilde{\alpha}(t_{m-1})$, already defined, lies in some V_β. Since V_β is *homeomorphic* to U_m by the map p, we define $\tilde{\alpha}(t)$ for $t_{m-1} \leq t \leq t_m$ to be the unique point of V_β so that $p \cdot \tilde{\alpha}(t) = \alpha(t)$. This is clearly continuous.

To show it is unique, note that any continuous map $\tilde{\alpha}$ from $[t_{m-1}, t_m]$ to the disjoint union

$$p^{-1}(U_m) = \bigcup_\beta V_\beta$$

which is prescribed at t_{m-1}, must send all of $[t_{m-1}, t_m]$ into a single V_β. (If not, take the inverse images of different V_β's and disconnect $[t_{m-1}, t_m]$.) But because V_β is homeomorphic to U by the map p, and because $p \cdot \tilde{\alpha}(t) = \alpha(t)$, $t_{m-1} \leq t \leq t_m$, $\tilde{\alpha}$ is specified uniquely by the formula $\tilde{\alpha}(t) =$ that unique point of V_β so that ρ sends it to $\alpha(t)$.

Now, to define $\tilde{\alpha}$ in total, to begin at $\tilde{\alpha}(0) = \tilde{x}_0$, and repeatedly apply the above to get a unique $\tilde{\alpha}$ as desired.

Remarks. This proposition is obviously technical, and somewhat tedious. Its importance is manifested by its repeated application below. We can use it to relate our work to fundamental groups and homotopy.

Proposition 9.2

Let $p: \tilde{X} \to X$ be a covering space, $p(\tilde{x}_0) = x_0$. Let α and β be homotopic maps from I to X, with common initial point x_0, which is held fixed during the homotopy. All these maps are assumed to be continuous.

Then the map $\tilde{\alpha}$ and $\tilde{\beta}$ (from Proposition 9.1) are homotopic, with initial points held fixed at $\tilde{x}_0$.

Proof. We have a map

$$F: I \times I \to X$$

so that

$$F(t, 0) = \alpha(t), \quad F(t, 1) = \beta(t)$$

$$F(0, u) = x_0.$$

Divide $I \times I$ into small rectangles, so that each rectangle lies entirely in $F^{-1}(U)$ for some connected, open U and we have the disjoint union

$$p^{-1}(U) = \bigcup_\alpha V_\alpha.$$

Define $\tilde{F}: I \times I \to \tilde{X}$, with

$$\tilde{F}(t, 0) = \tilde{\alpha}(t); \quad \tilde{F}(t, 1) = \tilde{\beta}(t)$$

$$\tilde{F}(0, u) = \tilde{x}_0$$

successively on the rectangles using the fact that each rectangle is connected so that if our map is defined in a part of a rectangle, to lie in one V_α, then it must send the entire rectangle into that V_α.

This is a straightforward generalization of the method of building $\tilde{\alpha}$ in Proposition 9.2.

The uniqueness of $\tilde{F}$ forces the conditions

$$\tilde{F}(t, 0) = \tilde{\alpha}(t) \quad \text{and} \quad \tilde{F}(t, 1) = \tilde{\beta}(t)$$

because these equations are certainly true when one applies p to either side, and they all share the common value $\tilde{x}_0$ when $t = 0$.

The reader should supply the details here.

Proposition 9.3

Let $p: \tilde{X} \to X$ be a covering space, with $p(\tilde{x}_0) = x_0$. Then the map

$$p_\#: \pi_1(\tilde{X}, \tilde{x}_0) \to \pi_1(X, x_0)$$

(see Theorem 8.1) is 1-1 (or, as this is often called, a monomorphism.)

Proof. Recall from the previous chapter that

$$p_\#(\{\alpha\}) = \{p \cdot \alpha\}.$$

Notice also that if $p_\#(\{\alpha\}) = p_\#(\{\beta\})$, then $p_\#(\{\alpha\}) \cdot p_\#(\{\beta\})^{-1} = e$, which is the same as

$$p_\#(\{\alpha\} \cdot \{\beta\}^{-1}) = e.$$

Thus, to show $p_\#$ is 1-1, we must show that $p_\#(\{\gamma\}) = e$ implies $\{\gamma\} = e$. (These remarks are basic material in any group theory text.)

But now, if $p_\#(\{\gamma\}) = e$, $p \cdot \gamma \simeq x_0$, where we write—by a slight abuse of terminology—x_0 for the constant path at x_0, with the homotopy keeping end points fixed, then, using Proposition 9.2, we see that the unique lifts of $p \cdot \gamma$ and x_0—that is the unique paths in $\tilde{X}$ which begin at $\tilde{x}_0$ and cover these paths—are homotopic, holding the initial point fixed.

Clearly, the constant path at $\tilde{x}_0$ is the unique lift of the constant path at x_0. As $p \cdot \gamma$ lifts to γ, the homotopy coming from Proposition 9.2 is indeed a homotopy of γ and $\tilde{x}_0$. But we must check that the initial *and* end points are held fixed at $\tilde{x}_0$ during the homotopy to complete the proof.

The initial point causes no trouble, as Proposition 9.2 yields at once that it is held fixed—that is $\tilde{F}(0, u) = \tilde{x}_0$, with $\tilde{F}: I \times I \to \tilde{X}$ being the homotopy. Suppose that the end point is not held fixed, i.e. the path $\tilde{F}(1, u)$ is a non-constant path. Since the homotopy $\tilde{F}$ lifts the homotopy of $p \cdot \gamma$ and x_0—from Proposition 9.2—and the homotopy of $p \cdot \gamma$ and x_0 leaves the end point fixed, $p \cdot \tilde{F}(1, u) = x_0$ for all $u \in I$.

If $\tilde{F}(1, u)$ is non-constant, and $p \cdot \tilde{F}(1, u) = x_0$, for all u, chose U, $x_0 \in U$, and a disjoint union

$$p^{-1}(U) = \bigcup_\alpha V_\alpha,$$

and we see that $\tilde{F}(1, u)$ has points in different V_α's. Indeed if all points were in one V_α and $p \cdot \tilde{F}(1, u) = x_0$, then $\tilde{F}(1, u)$ would be constant.

Selecting disjoint open sets in $p^{-1}(U)$, both of which contain points of $\tilde{F}(1, u)$, and taking their inverse images under the path in question, we disconnect I for a contradiction. Hence $\tilde{F}(1, u)$ is constant, which means that the end point is fixed during the homotopy. This completes the proof.

Remark. One might regard this proposition as a necessary condition for a space to be covering space of X. More importantly, it is a first step in the classification of covering spaces over a fixed space (see Theorem 9.1). There is one element of vagueness in Propositions 9.1, 9.2 and 9.3. That is we have chosen $\tilde{x}_0 \in p^{-1}(x_0)$. The selection of different $\tilde{x}_0$ obviously yields different lifts for a given path (see examples at beginning). But in case of the last proposition, we can say more. Therefore, we detour for a moment to clear this up.

Proposition 9.4

Let $\tilde{x}_0$ and $\tilde{x}_0'$ be points in $\tilde{X}$, a covering space of X. Suppose $p(\tilde{x}_0) = p(\tilde{x}_0') = x_0$. Then the two subgroups of $\pi_1(X, x_0)$, namely

$$p_\#(\pi_1(\tilde{X}, \tilde{x}_0)) = \{\{\gamma\} \mid \{\gamma\} \in \pi_1(X, x_0), \{\gamma\} = p_\#(\{\alpha\}), \{\alpha\} \in \pi_1(\tilde{X}, \tilde{x}_0)\}$$

and

$$p_\#(\pi_1(\tilde{X}, \tilde{x}_0')) = \{\{\gamma\} \mid \{\gamma\} = p_\#(\{\beta\}), \{\beta\} \in \pi_1(\tilde{X}, \tilde{x}_0')\}$$

and conjugate. That is, there is $\{\tau\} \in \pi_1(X, x_0)$ such that

$$\{\tau\} \, (p_\# \pi_1(\tilde{X}, \tilde{x}_0)) \, \{\tau\}^{-1} \equiv p_\# \pi_1(\tilde{X}, \tilde{x}_0').$$

In other words, change in base point over x_0 yields conjugate subgroups.

We also have the converse: given such $\{\tau\}$, then there is a new base point in $\tilde{X}$ so that the image of the fundamental group by $p_\#$ is the conjugate, of the image of the group with old base point, by this element $\{\tau\}$.

Proof. Let $\tilde{\tau}: I \to \tilde{X}$ with

$$\tilde{\tau}(0) = \tilde{x}_0', \quad \tilde{\tau}(1) = \tilde{x}_0.$$

Then $p \cdot \tilde{\tau}$ is a closed path, with end points at x_0, in X. Put $\tilde{\tau}^{-1}(t) = \tilde{\tau}(1-t)$, and $\tau = p \cdot \tilde{\tau}$.

Then $(\tilde{\tau} * \alpha) * \tilde{\tau}^{-1}$, for α a closed path based at $\tilde{x}_0$, is then a closed path based at $\tilde{x}_0'$. This implies at once, that the conjugate of an element of $p_\#(\pi_1(\tilde{X}, \tilde{x}_0))$ is an element of $p_\#(\pi_1(\tilde{X}, \tilde{x}_0'))$. Conjugation is clearly a 1-1 map of elements, so we must prove that every element of $p_\#(\pi_1(\tilde{X}, \tilde{x}_0'))$ is of the desired form. Given an element $\{\beta\} \in \pi_1(\tilde{X}, \tilde{x}_0')$, consider

$$[([(\tilde{\tau} * \tilde{\tau}^{-1}) * \beta] * \tilde{\tau}) * \tilde{\tau}^{-1}.]$$

This visibly represents — after projection by $p_\#$ — the element

$$\{\tau\} \cdot \{\tau\}^{-1} \cdot \{\beta\} \cdot \{\tau\} \cdot \{\tau\}^{-1} = \{\beta\},$$

where $\tau = p \cdot \tilde{\tau}$. But putting $\alpha = (\tilde{\tau}^{-1} * \beta) * \tilde{\tau}$,

$$\{\beta\} = \{\tau\} \cdot \{\alpha\} \cdot \{\tau\}^{-1}.$$

Thus every element of $p_\#(\pi_1(\tilde{X}, x_0'))$ is of the desired form and the subgroups $\{\tau\}(p_\# \pi_1(\tilde{X}, \tilde{x}_0))\{\tau\}^{-1}$ and $p_\#(\pi_1(\tilde{X}, \tilde{x}_0'))$ are equal.

To prove the final assertion, choose from Proposition 9.1 a path $\tilde{\tau}$ beginning at the old base point and lying over the closed path τ in X. The end point of $\tilde{\tau}$ is trivially checked to be the desired new base point so that the image, under $p_\#$, of the fundamental group, with this new base point, is obtained from the image, with the old base point, by conjugation with $\{\tau\}$. (Check details here!)

The various propositions above yield the following corollary.

Corollary 9.1

The cardinality (or number of elements in) the set $p^{-1}(x_0)$ is the index of the subgroup

$$p_\#(\pi_1(\tilde{X}, \tilde{x}_0))$$

in the group $\pi_1(X, x_0)$. This index (and hence the cardinality of $p^{-1}(x_0)$) is independent of x_0, that is a constant for the covering space.

Proof. Given $\tilde{x}_0 \in p^{-1}(x_0)$ fixed, choose another $\tilde{x}_0'$; as in proof of Proposition 9.4, choose $\tilde{\tau}$. If $\{\alpha\} \in \pi_1(\tilde{X}, \tilde{x}_0)$

$$p \cdot (\alpha * \tilde{\tau}^{-1})$$

is a closed loop in X based at x_0 which clearly represents

$$(p_\#(\{\alpha\}) \cdot \{\tau^{-1}\}) \in \pi_1(X, x_0)$$

with $\tau = p \cdot \tilde{\tau}$.

This is directly checked—as in the above proposition—to be a 1-1 correspondence of $p^{-1}(x_0)$ to the cosets

$$(p_\#(\pi_1(\tilde{X}, \tilde{x}_0)))\{\tau^{-1}\}.$$

For example, if we choose another $\tilde{\tau}'$, then

$$(\tilde{\tau}')^{-1} * (\tilde{\tau})$$

is a closed loop at $\tilde{x}_0$, so that $\{\tau\}$ and $\{\tau'\}$ have the property that $\{\tau'\}^{-1} \cdot \{\tau\}$ $\in p_\#(\pi_1(\tilde{X}, \tilde{x}_0))$. But then they define the same coset

$$p_\#(\pi_1(\tilde{X}, \tilde{x}_0)) \cdot \{\tau^{-1}\} \equiv p_\#(\pi_1(\tilde{X}, \tilde{x}_0)) \cdot \{(\tau')^{-1}\}$$

(see any basic group theory text). The rest of the verification that this is a 1-1 correspondence is left to the reader to check. (Do it!)

Finally, we need that the index is constant in x. Let x_1 be another point of X, $\sigma: I \to X$ a path from x_0 to x_1; σ lifts to $\tilde{\sigma}$ from $\tilde{x}_0$ to $\tilde{x}_1$. Using the Proposition 8.2 on effect of change of base point, we see that the action of σ takes $p_\#(\pi_1(\tilde{X}, \tilde{x}_0))$ to $p_\#(\pi_1(\tilde{X}, \tilde{x}_1))$, in fact it is the same to let $\tilde{\sigma}$ act and then apply $p_\#$, or to apply $p_\#$ and then let σ act.

We then define the action of σ on a coset

$$(p_*(\pi_1(\tilde{X}, \tilde{x}_0))) \cdot \{\tau^{-1}\}$$

to be

$$(p_*(\pi_1(\tilde{X}, \tilde{x}_1))) \cdot \phi_\sigma(\{\tau^{-1}\}).$$

This map has an obvious inverse, namely, the result of the same process for the path σ^{-1} defined by $\sigma^{-1}(t) = \sigma(1 - t)$. Hence, it is a one-to-one correspondence of the cosets of $p_*(\pi_1(\tilde{X}, \tilde{x}_0))$ and $(p_*(\pi_1(\tilde{X}, \tilde{x}_1)))$, completing the proof.

The cardinality of $p^{-1}(x_0)$ is called the *number of sheets* of the covering space.

We now pause to illustrate these theorems in the case of an example. Consider Example 3(c) of the covering space

$$p \colon \mathbb{R} \to S^1.$$

Let $\alpha \colon I \to S^1$ be a path which, for example, we suppose wraps around S^1 two times. To find a lifting of α to $\mathbb{R}$, we proceed along pieces U of S^1 over which $p^{-1}(U)$ splits as a disjoint union of open sets. Suppose we wish to find $\tilde{\alpha} \cdot I \to \mathbb{R}$ so that $p \cdot \tilde{\alpha} = \alpha$ and $\tilde{\alpha}(0) = 1$. We easily calculate that $\tilde{\alpha}$ runs linearly from 1 to 3, or $\tilde{\alpha}(t) = 2t + 1$. This is illustrated as follows:

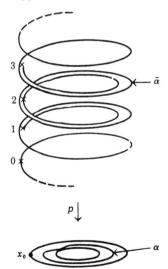

Note that because of our definitions of p, the scale in $\mathbb{R}$ is different from that in S^1 and that we have chosen a direction in S^1.

Since $\mathbb{R}$ is contractible (see Chapter 8), $\pi_1(\mathbb{R}, \tilde{x}_0)$ is the trivial group.

166 Topology

$\pi_1(S^1, x_0)$ is the group of integers (proof in next chapter or "Problems" at the end of this chapter). $p_\#$ is the obvious monomorphism (1-1 map) of the trivial group, with only the identity element, to the integers, i.e. $p_\#(e) = 0$.

Before proceeding to the classification theorem, it is useful to call attention to a result which shows how a geometric problem may be resolved entirely in terms of algebra. This proposition is a very clear example of the general philosophy of algebraic topology, which is to reduce hard geometric problems to more manageable algebraic ones.

Proposition 9.5

Let $p \colon \tilde{X} \to X$ be a covering space. Suppose Y is arcwise connected and locally arcwise connected, that is, Y is subjected to the same point-set conditions as X and $\tilde{X}$. Let $f \colon Y \to X$ be a continuous map. Suppose $y_0 \in Y$ with $f(y_0) = x_0 \in X$, and $\tilde{x}_0 \in \tilde{X}$ with $p(\tilde{x}_0) = x_0$.

Then a necessary and sufficient condition that f lift to $\tilde{f} \colon Y \to \tilde{X}$ with $\tilde{f}(y_0) = \tilde{x}_0$ and $p \cdot \tilde{f}(y) = f(y)$ for all $y \in Y$ is

$$f_\#(\pi_1(Y; y_0)) \subseteq p_\#(\pi_1(\tilde{X}, \tilde{x}_0))$$

(recall that if $h \colon G_1 \to G_2$ is a homomorphism of groups, $h(G_1)$ means $\{u \mid u \in G_2, u = h(v), \text{ for some } v \in G_1\}$).

Proof. Suppose that f lifts to $\tilde{f}$. Then we have a commutative diagram

so that the two homomorphisms $f_\#$ and $p_\# \cdot \tilde{f}_\#$, from $\pi_1(Y, y_0)$ to $\pi_1(X, x_0)$ will be the same.

But then

$$f_\#(\{\alpha\}) = p_\#(\tilde{f}_\#(\{\alpha\})) \subseteq p_\#(\pi_1(\tilde{X}, x_0)).$$

Conversely, suppose the condition. Consider $y \in Y, f(y) \in X$, and choose a path $\sigma \colon I \to Y$ so that

$$\sigma(0) = y_0, \quad \sigma(1) = y$$

Then $f \cdot \sigma \colon I \to X$ is a path from x_0 to $f(y)$. Define $\widetilde{(f \cdot \sigma)}$ to be the lift to $\tilde{X}$ guaranteed by Proposition 9.1, i.e. $\widetilde{(f \cdot \sigma)}$ is that unique path in $\tilde{X}$ lying over the path $f \cdot \sigma$ in X. We then define

$$\tilde{f}(y) = \widetilde{(f \cdot \sigma)}(1).$$

That is, we let $\tilde{f}(y)$ be the end point of this lifted path.

There is an apparent danger here, that $\tilde{f}$ is ill defined, for although the lifting is quite unique, the original choice of the path σ in Y is not at all unique. Let $\bar{\sigma}$ be another such path and $\bar{\sigma}^{-1}(t) = \bar{\sigma}(1-t)$ as usual. Then $\bar{\sigma}*\sigma^{-1}$ is a closed loop, based at y_0, in the space Y. Our assumption means that

$$f_\#(\{\bar{\sigma}*\sigma^{-1}\})$$

is in the image of $p_\#$, i.e. for some $\alpha: I \to \tilde{X}$, $\alpha(0) = \alpha(1) = \tilde{x}_0$, we have

$$f_\#(\bar{\sigma}*\sigma^{-1}) = p_\#(\{\alpha\}).$$

In other words,

$$f \cdot (\bar{\sigma}*\sigma^{-1}) \simeq p \cdot \alpha$$

with end points fixed. Using Proposition 9.2, this homotopy may be lifted to a homotopy of maps into $\tilde{X}$, rather than X. And as liftings for paths are unique by Proposition 9.1, the end of the homotopy must be α (check this). Denote the beginning of the homotopy, the map which is the lift of $f \cdot (\bar{\sigma}*\sigma^{-1})$, by $\omega: I \to \tilde{X}$, and we must have

$$\omega(0) = \omega(1) = \tilde{x}_0;$$
$$p \cdot \omega = f \cdot (\bar{\sigma}*\sigma^{-1}).$$

But $f \cdot (\bar{\sigma}*\sigma^{-1})$ runs through $f \cdot \bar{\sigma}$ in the first half of the interval, and $f \cdot \sigma^{-1}$ in the second half; hence the value $\omega(\frac{1}{2})$, which lies over $f \cdot (\bar{\sigma}*\sigma^{-1})(\frac{1}{2})$, is at the same time $(\widetilde{f \cdot \sigma})(1)$ and $(\widetilde{f \cdot \bar{\sigma}})(1)$. (Check this!) Thus, $\tilde{f}(y)$ is well-defined, no difficulties arising from the choice of the path.

It remains to prove $\tilde{f}$ is continuous. Take an open $U \subseteq X$, so that, by our definition of a covering space, we have

$$p^{-1}(U) = \bigcup_\alpha V_\alpha.$$

Let $y \in f^{-1}(U) \subseteq Y$, where continuity of f assures that $f^{-1}(U)$ is open. Since Y is locally arcwise connected, we may find an arcwise connected open set $W \subseteq Y$ so that

$$y \in W \subseteq f^{-1}(U).$$

Now suppose $\tilde{f}(y) \in V_\alpha$, with $\tilde{f}(y) = (\widetilde{f \cdot \sigma})(1)$. If $y_1 \in W$ is another point choose a path τ from y to y_1 in W and set

$$\sigma_1 = \sigma*\tau$$

and thus $f \cdot \sigma_1 = f \cdot (\sigma*\tau) = (f\sigma)*(f\tau)$. (Check!) Since $f \cdot \sigma$ lifts to $(\widetilde{f \cdot \sigma})$, which ends at $\tilde{f}(y)$, and $f \cdot \tau$ lies entirely in U, so that its lift lies entirely within V_α, we conclude that $\tilde{f}(y_1) \in V_\alpha$ for every $y_1 \in W$.

The conclusion is then that for this $y \in Y$ there is an open set W, $y \in W$, so that $\tilde{f}(W) \subseteq V_\alpha$. This being the case for any $y \in \tilde{f}^{-1}(V_\alpha) \subseteq f^{-1}(U)$, we see that $\tilde{f}^{-1}(V_\alpha)$ is open. But I claim that the V_α—for all such possible

U—form a basis for the topology at $\tilde{X}$. After we check this, the proof will be complete.

To this end, let $0 \subseteq \tilde{X}$ be open. Since all sufficiently small open sets in $\tilde{X}$, that is, those lying inside some V_α, are sent by p to open sets in X, we see that $p(0)$ must be open. If $\tilde{x} \in 0$, $p(\tilde{x}) = x$ belongs to $p(0)$. By the definition of a covering space, there is U open, with $x \in U$, with

$$U \subseteq p(0)$$

and

$$p^{-1}(U) = \bigcup_\alpha V_\alpha.$$

For some α, $\tilde{x} \in V_\alpha \cap 0$. Then $p(V_\alpha \cap 0)$ is an open set, smaller than U, so that it, too, is covered by a disjoint union of homeomorphic copies of itself. (It is easy to see that we may assume, with no loss of generality, that it is connected.) If we write

$$p^{-1}(p(V_\alpha \cap 0)) = \bigcup_\beta V_\beta^1.$$

Then for some V_β^1, $\tilde{x} \in V_\beta^1 \subseteq V_\alpha \cap 0 \subseteq 0$.

This immediately shows that 0 is the union of such V_β^1, and that these sets form a basis, polishing off the proof.

Remarks. (We quote some facts from Chapter 10 as illustrations here.)

1) Suppose we know $\pi_1(\mathbb{R}, r_0) = 0$—the trivial group, while $\pi_1(S^1, x_0) = \mathbb{Z}$—the group of integers. Then we know that the identity map $S^1 \to S^1$ does not lift to a map $\tilde{1}\colon S^1 \to \mathbb{R}$ with

$$p \cdot \tilde{1} = 1\colon S^1 \to S^1.$$

This trivial application can be checked by elementary means because any map of the compact S^1 to $\mathbb{R}^1$ has a maximum. In a neighborhood of such a maximum, it is not possible for a map to cover the identity on S^1. (Draw a picture.)

2) Let $p\colon S^1 \to S^1$ be the covering space (Example 3(b) above) which winds S^1 around itself 5 times. Let $f\colon S^1 \to S^1$ wind around 3 times, i.e. $f(e^{i\theta}) = e^{3i\theta}$. Then f does *not* lift to $\tilde{f}$ because the integers which are divisible by 3 do *not* form a subgroup of those divisible by 5. In fact, we shall see (in the next chapter) that the map f has the effect of multiplying by 3, p by 5, etc.

3) If $\pi_1(Y, y_0) = 0$ (the trivial group), then every f lifts to $\tilde{f}\colon Y \to \tilde{X}$.

If we were to apply this to $Y = I$, we would recover Proposition 9.1 (unfortunately we have already used it in the proof of Proposition 9.5).

4) Note that whenever $\pi_1(X, x_0) = 0$, X can have no covering space other than itself. For if $p\colon \tilde{X} \to X$ were one, we apply the above Proposition 9.5 to $1\colon X \to X$, and get

$$s\colon X \to \tilde{X}$$

with

$$p \cdot s = 1.$$

Suppose $p^{-1}(x_0)$ has at least 2-points. Let σ be a path from $\tilde{x}_0$ to another. Then $s \cdot p \cdot \sigma$ is a path which lies over the same path as σ in X, namely $p \cdot \sigma$, for $p(s \cdot p \cdot \sigma) = (p \cdot s) p \cdot \sigma = p \cdot \sigma$.
But $s \cdot p \cdot \sigma = s \cdot (p \cdot \sigma)$ is the result of applying the continuous s to a closed path $p \cdot \sigma$. By uniqueness of lifting (Proposition 9.1) σ and $s \cdot p \cdot \sigma$ (which have the same image by p) are the same; thus σ must be closed. This immediately proves that $p^{-1}(x_0)$ has one element, or $p: \tilde{X} \to X$ is a homeomorphism (globally).

Easy examples are the Euclidean spaces $\mathbb{R}^n$, which admit no non-trivial covering spaces.

5) If $p: \tilde{X} \to X$ is a non-trivial covering space, and $\pi_1(\tilde{X}, x_0) = 0$ (for example $\rho: \mathbb{R} \to S^1$), then there is always a map $S^1 \to X$ which does not lift to a map $S^1 \to \tilde{X}$. To see this, note

a) $\pi_1(X, x_0) \neq 0$ [or else, by 4) above, we could not even have a non-trivial covering space].

b) Any map $f: S^1 \to X$, which represents a non-zero element of $\pi_1(X, x_0)$, will send $1 \in \pi_1(X^1, x_0) = Z$ to this non-zero element (immediate from computation in next chapter). Since $p_\#$ is the zero homomorphism, i.e. sends everything to e, because its domain is trivial, the map f must violate the condition of Proposition 9.5. if it lifts to $\tilde{X}$.

We conclude that whenever $p: \tilde{X} \to X$ is non-trivial but $\pi_1(\tilde{X}, x_0)$ is trivial, we have a loop in X over which the covering space is not just a disjoint union of copies of the loop.

More generally, these remarks are valid whenever $p: \tilde{X} \to X$ in non-trivial, even if $\pi_1(\tilde{X}, x_0) \neq 0$, for then there will always be elements in $\pi_1(X, x_0)$ not in the image of $p_\#$, but which by b) above may be taken in the image of $f_\#$.

6) A map $f: \tilde{X} \to \tilde{X}$, where $p: \tilde{X} \to X$ is a covering space, is called a deck transformation if

$$p \cdot f = p$$

or equivalently

is a commutative diagram.

The set of deck transformations forms a group under composition. The group acts as a group of transformations on $p^{-1}(x_0)$, for any $x_0 \in X$. The group of deck transformations is an interesting topic, but one which would lead us far afield. More information can be found in the book by W. S. Massey.

We can now tackle the first major theorem, which discusses the uniqueness of covering spaces.

Theorem 9.1

Let $p_1\colon \tilde{X}_1 \to X$ and $p_2\colon \tilde{X}_2 \to X$ be two covering spaces of X. (Assume $p_1(\tilde{x}_1) = p_2(\tilde{x}_2) = x_0$.) Suppose
$$(p_1)_\#(\pi_1(\tilde{X}_1, \tilde{x}_1)) \equiv (p_2)_\#(\pi_1(\tilde{X}_2, \tilde{x}_2)).$$
That is, the images of the fundamental groups of $\tilde{X}_1$ and $\tilde{X}_2$, in the fundamental group of X, are identical.

Then there is a homeomorphism $f\colon \tilde{X}_1 \to \tilde{X}_2$ such that $p_2 \cdot f = p_1$, which preserves base points.

Proof. Let p_2 play the role of the covering space in Proposition 9.5, and let p_1 play the role of f. Then our condition assures that we have
$$f\colon \tilde{X}_1 \to \tilde{X}_2$$
with $p_2 \cdot f = p_1$.

Reversing the roles of $\tilde{X}_1$ and $\tilde{X}_2$, we obtain
$$g\colon \tilde{X}_2 \to \tilde{X}_1.$$
The proof will be complete, provided that we can show that the two compositions $g \cdot f$ and $f \cdot g$ are the identity maps. For this, it suffices to show that any map $h\colon \tilde{X}_1 \to \tilde{X}_1$, with $h(\tilde{x}_1) = \tilde{x}_1$ and $p_1 \cdot h = p_1$, is the identity (for the same will hold for $\tilde{X}_2$).

Let h be such a map, and take $x \in \tilde{X}_1$. Let $\tilde{x}_1$ be the base point and let
$$\alpha\colon I \to \tilde{X}_1,$$
be a path from $\tilde{x}_1$ to x. Then our conditions on h immediately imply that the paths α and $h \cdot \alpha$ (defined by $h \cdot \alpha(t) = h(\alpha(t))$) both begin at $\tilde{x}_1$ and both cover the path $p_1 \cdot \alpha$. But then the uniqueness assertion of Proposition 9.1 guarantees that α and $h \cdot \alpha$ are the same path; they then have the same end points, namely x and $h(x)$.

We have shown that $x = h(x)$ for all x, proving that h is the identity and completing our theorem.

Remarks. Theorem 9.1 may be generalized in two ways. First, if the groups $(p_1)_\#(\pi_1(\tilde{X}_1, \tilde{x}_1))$ and $(p_2)(\pi_1(\tilde{X}_2, \tilde{x}_2))$, both subgroups of $\pi_1(X, x_0)$ are not equal, but are conjugate, then the covering spaces are equivalent in the sense that there is a homeomorphism f which does not preserve the base points.

Secondly, if $(p_1)_\#(\pi_1(\tilde{X}_1, \tilde{x}_1)) \subseteq (p_2)_\#(\pi_1(\tilde{X}_2, \tilde{x}_2))$ then the lifting map

$p_1 \colon \tilde{X}_1 \to \tilde{X}_2$ guaranteed by Proposition 9.5 is actually a covering map (in this case, where f is replaced by the rather special map p_1). For details, we refer again to W. S. Massey's book (see Bibliography).

We now wish to tackle the existence of covering spaces. Our work is motivated by a simple observation. If $p \colon \tilde{X} \to X$ is a covering space, $U \subseteq X$ arcwise connected and

$$p^{-1}(U) = \bigcup_\alpha V_\alpha$$

with the V_α disjoint and mapped homeomorphically onto U by p, then each V_α may be described in the following somewhat complex way:

Choose x_0 and $\tilde{x}_0$ with $p(\tilde{x}_0) = x_0$. Suppose $x \in U$, $\tilde{x} \in V_\alpha$ with $p(\tilde{x}) = x$. Select a path from x_0 to x that lifts to a path from $\tilde{x}_0$ to $\tilde{x}$, calling it β. Let $\{\gamma\}$ run over the set of all paths in V_α which begin at $\tilde{x}$ and of course end in V_α. Then the composite path $\beta * \gamma$ begins at $\tilde{x}_0$ and ends at some point in V_α. As V_α is arcwise connected, the set of all end points of all such paths $\beta * \gamma$ fills up V_α. This elaborate description of V_α suggests how to build V_α's, and hence $\tilde{X}$, in our existence theorem.

Note that if we select another path in X, such that composed with our original we have a closed loop, which represents an element in $p_\#(\pi_1(\tilde{X}, \tilde{x}))$, then the lift of the two paths has a common end point, namely $\tilde{x}$. This is the extent to which these remarks are free from the choices we have made.

Our goal shall be to reverse these ideas to fabricate $\tilde{X}$ from any subgroup of $\pi_1(X, x_0)$. Unfortunately, we need a definition. This will probably appear rather technical, but we shall soon see that it is exactly what is needed to ensure that the construction in our main existence theorem works out; our first definition is a well known term; the more general second definition is what we need later.

Definition 9.3

Let X be an arcwise connected space, $x_0 \in X$
a) X is *simply connected*, if $\pi_1(X, x_0)$ is the trivial group, that is contains the single element e.
b) X is *semi-locally simply connected*, if given $x \in X$ and an open U, $x \in U$, there is an open V, which is arcwise connected,

$$x \in V \subseteq U$$

so that if $y \in V$ and α and β are two paths in V, both beginning at x and ending at y, *then* α may be deformed into β by a deformation or homotopy which leaves the beginning and end points fixed, but which may possibly, at some time during the homotopy, pass out of V.

Remarks. 1) Contractible spaces are simply connected. Simply connected spaces are easily shown to be semi-locally simply connected (though we don't use this fact).

2) Manifolds (such as S^1) need not be simply connected, but are always semi-locally simply connected. In fact, every point has an open neighborhood, which is homeomorphic to $\mathbb{R}^n$. But in such a neighborhood, it is trivial to deform a path α into a path β. In this case, the deformation actually remains within the neighborhood. The term "semi-locally" refers to the possibility that the deformation pass outside of the small set.

3) A deeper, more instructive example is the following. Consider any infinite sequence of circles, all tangent at a point, and whose radii approach zero.

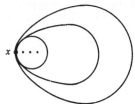

Erect a cone over this planar figure in space, i.e.

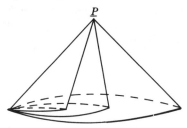

This means that we take all lines from a point $\underline{P}$ to the figure, and give the resulting set the relative topology of a subset of Euclidean space $\mathbb{R}^3$.

I claim that this space is semi-locally simply connected, but where the deformation required in the definition may pass out of the set U. Choose a small open ball about x, small enough to avoid $\underline{P}$. Select y to be diametrically opposed to x on some small circle which lies entirely within U.

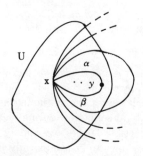

(The region about x describes the points of U, as those points of our space which lie entirely within the region.) α and β are the "top" and "bottom" paths from x to y. It is easy to check that if α and β may be deformed into one another, then the map which wraps I around S^1 once is null-homotopic and this easily implies $\pi_1(S^1, x)$ is trivial (contradicting the fact, proved in Chapter 10, and also sketched in the "Problems" at the end of this chapter, that $\pi_1(S, x) = Z$, the integers).

But α and β may indeed be deformed to one another in the entire space, by lifting either path over the point P and then deforming it back down to the other path. This is vaguely sketched as follows, where $\beta_1, \beta_2, \cdots$ shows stages of the deformation

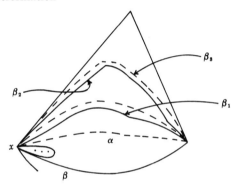

Thus, we have again a semi-locally simply connected space (check that the condition of Definition 9.3(b) is trivially satisfied for open neighborhoods of other points than 0).

Theorem 9.2

Let X be arcwise connected and semi-locally simply connected (thus locally arcwise connected). Let $x_0 \in X$ and

$$H \subseteq \pi_1(X, x_0)$$

be a subgroup.

Then there exists a covering space

$$p \colon \tilde{X} \to X$$

such that

$$p_\#(\pi_1(\tilde{X}, \tilde{x}_0)) = H \subseteq \pi_1(X, x_0)$$

(for $\tilde{x}_0 \in p^{-1}(x_0)$). As $p_\#$ is 1-1 (see Proposition 9.3), we conclude that $\pi_1(\tilde{X}, \tilde{x}_0)$ is isomorphic to H.

Proof. Let $\mathcal{P}$ be the set of all paths in X beginning at x_0. If α and β are two paths, beginning at x_0 and ending at x, define an equivalence relation $\alpha \sim \beta$ if the path $\alpha*\beta^{-1}$ (where $\beta^{-1}(t) = \beta(1-t)$) represents an element in the subgroup H. (Check that this is an equivalence relation.)

We set $\tilde{X} = \mathcal{P}/\sim$; $p\colon \tilde{X} \to X$ is $p(\{\alpha\}) = \alpha(1)$.

Let U be an open set, $\alpha\colon I \to X$ be a path, in X, which begins at x_0 and ends at a point of U, say x. Define the set (U, α) to be the set of equivalence classes of paths $\alpha*\beta$, where β is a path that begins at x and lies entirely within U. Clearly (U, α) does not depend on the equivalence class of α.

These sets (U, α) with U satisfying our local conditions will build the topology on $\tilde{X}$. Note that if $\{\beta\} \in (U, \alpha)$, then $(U, \beta) = (U, \alpha)$. For suppose $\beta \sim \alpha*\gamma$. Then $\beta*\delta \sim \alpha*(\gamma*\delta)$, showing that any point of (U, β) is also a point of (U, α). On the other hand, α is clearly equivalent to a path $\beta*\sigma$, because $\beta*\gamma^{-1} \sim \alpha*(\gamma*\gamma^{-1})$ and end point preserving homotopies, such as that between $\gamma*\gamma^{-1}$ and the constant map, keep within equivalence classes. But then any point of (U, α), by the same argument as above, must belong to (U, β). We then may deduce that whenever $\alpha(1)$ and $\beta(1)$ belong to U, then either $(U, \alpha) \equiv (U, \beta)$, or they are disjoint.

Suppose $\{\alpha\} \in (U, \beta) \cap (V, \gamma)$. Our earlier remarks show that

$$(U, \beta) = (U, \alpha)$$

and

$$(V, \gamma) = (V, \alpha).$$

Then $\{\alpha\} \in (U, \alpha) \cap (V, \alpha)$ with $\alpha(1) \in U \cap V$. It is then immediate that

$$\{\alpha\} \in (U \cap V, \alpha) \subseteq (U, \alpha) \cap (V, \alpha) \equiv (U, \beta) \cap (V, \gamma).$$

It is then clear that these sets (U, α) form a basis for a topology on $\tilde{X}$, and we define the topology on $\tilde{X}$ in terms of this basis.

I claim that $p\colon \tilde{X} \to X$, defined above as $p(\{\alpha\}) = \alpha(1)$, is both onto and continuous. If $x \in X$, choose $\alpha\colon I \to X$ beginning at x_0 and ending at x. Then, $p(\{\alpha\}) = \alpha(1) = x$, showing p is onto. On the other hand, if $U \subseteq X$ is open, and $\{\alpha\}$ lies over U, that is $p(\{\alpha\}) \in U$, then (U, α) is an open set in $\tilde{X}$ with $p((U, \alpha)) \subseteq U$. Thus, for every element in $p^{-1}(U)$, we have found an open set, containing that element and lying in $p^{-1}(U)$. $p^{-1}(U)$ is visibly the union of all these open sets, and hence, it is open. This shows that p is continuous.

We must now establish the decomposition

$$p^{-1}(U) = \bigcup_\alpha V_\alpha$$

where the V_α are disjoint open sets, each mapped homeomorphically onto U, by the map p. Choose U as V in the definition of semi-locally simply connected. One may show, without difficulty, that the collection of all such

U forms a basis for the topology of X. It is clear that if (W, α), with W any open set, is as above, and U is as V in the definition of a semi-locally simply connected set, containing $\alpha(1)$ and lying in W, then we have at once

$$(U, \alpha) \subseteq (W, \alpha).$$

Thus, the sets (U, α) for such U also form a basis for the topology on $\tilde{X}$.

By earlier remarks, $p^{-1}(U)$ is the disjoint union of (U, α_k) for some suitable set of α_k, all with $\alpha_k(1) \in U$. Consider the restriction of p to (U, α_k), namely $p \mid (U, \alpha_k) : (U, \alpha_k) \to U$. It is obviously continuous and onto. Let (V, β) be an open set in (U, α_k), which we may assume, by the previous paragraph is arcwise connected (check!). Clearly $p \mid (U, \alpha_k)$ maps (V, β) onto V, proving that V is open.

Lastly, we need that $p \mid (U, \alpha_k)$ is 1-1. This shall make strong use of our condition in Definition 9.3(b). Suppose we have

$$\{\beta_1\}, \{\beta_2\} \in (U, \alpha)$$

with $\beta_1(1) = \beta_2(1)$, that is to say, they have the same image under projection to X. By earlier remarks, $(U, \alpha) = (U, \beta_1) = (U, \beta_2)$. Therefore, $\beta_1 \sim \alpha * \gamma_1$ and $\beta_2 \sim \alpha * \gamma_2$. By the semi-local simply connected condition, the paths in question γ_1 and γ_2 are homotopic holding end points fixed. This shows at once that $\beta_1 \sim \beta_2$, or that $p \mid (U, \alpha_k)$ is 1-1. Thus, $p \mid (U, \alpha_k)$ is a homeomorphism of the open set (U, α_k) onto U.

There are left three points to showing that $\tilde{X}$ is covering space of X. That is that $\tilde{X}$ is a locally arcwise connected and arcwise connected Hausdorff space. The first and last of these three conditions are obvious from the above (check!), but we prove $\tilde{X}$ is arcwise connected. Let $\{\alpha\} \in \tilde{X}$. Define

$$\phi: I \to \tilde{X}$$

by $\phi(t)$ = the path which sends s to $\{\alpha(t \cdot s)\}$. We need to examine $\phi^{-1}((U, \alpha_1))$ which is the set of $t \in I$ so that $\alpha(t \cdot s)$ is equivalent to a path which has the form $\alpha_1 * \beta$, β lying entirely within U. But if for some fixed t_1, $\phi(t_1)$ lies in (U, α_1), then clearly all those points t which lie near enough to t_1 so that $\phi(t)$ ends at a point of U, will be an open neighborhood of t_1 which is mapped, by ϕ, to a subset of (U, α_1). This shows at once that ϕ is continuous. In fact, $\phi(t)$ just consists of a partial path along the route of α; you should draw a picture. $\tilde{x}_0$ is the constant path at x_0.

The reader should check the final fact that $p_*(\pi_1(\tilde{X}, \tilde{x}_0)) = H$, completing the proof. ($p_*(\pi_1(\tilde{X}, \tilde{x}_0))$ is represented by those closed paths in X, which lift to closed paths in $\tilde{X}$, or equivalently, two paths beginning at x_0 and ending at a point x, which lift to two paths which have the same end points over x. But $\alpha(s)$ is covered by $\alpha(t \cdot s)$, etc. Hence, $\alpha(t \cdot s)$ and $\beta(t \cdot s)$ will have the same end point, when $\{\alpha * \beta^{-1}\} \in H$.)

176 Topology

Corollary 9.2

a) If X satisfies the conditions of the Theorem 9.2, there is a covering space $\tilde{X}$ which is simply connected.

b) If Y satisfies these conditions and is simply connected, Y admits no covering space other than itself.

Proof. For a), take H to be the trivial subgroup in the theorem. If, for b), Y had a non-trivial covering space, the trivial group would have a proper subgroup of finite index (see Corollary 9.1).

Remark. 1) The unique (by Theorem 9.1) simply connected covering space of an arcwise connected, locally arcwise connected, semi-locally simply connected space X is called the *universal covering space* of X.

2) The number of sheets of the universal covering space equals the number of elements in—or the order of—the fundamental group $\pi_1(X, x_0)$, because the order of a group equals the index of the trivial subgroup.

3) The two Theorems 9.1 and 9.2 give an existence and uniqueness theory of covering spaces which is mirrored perfectly in the subgroup structure of the fundamental group. The student familiar with Galois theory will notice some similarities with this theory.

We may now draw various corollaries, to our basic theorems, which constitute computations of various fundamental groups. The first is an ingenious trick due to L. Smith.

Corollary 9.3

If $n > 1$, $\pi_1(S^n; x_0)$ is trivial.

Proof. S^n, the n-sphere, clearly meets the conditions of Theorem 9.2. Let $\tilde{S}^n$ be its universal covering space. Let D^n be the n-disc, i.e.

$$D^n = \{x \mid x \in \mathbb{R}^n, d(x, 0) \leq 1\}.$$

The boundary of D^n, say ∂D^n, is S^{n-1}.

Define

$$f: D^n \to S^n$$

to identify all of $\partial D^n = S^{n-1}$ to a point, say x_0. By Proposition 9.5 (using the contractible space $D^n = Y$), we see that f lifts to $\tilde{f}$, so that we have a commutative diagram

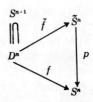

Because $n > 1$, S^{n-1} is connected. Because f maps all of S^{n-1} to the single point x_0, $\tilde{f}$ maps the connected space S^{n-1} to a disjoint union of points. However, as the image of a connected space is connected, we must have that $\tilde{f}$ maps S^{n-1} to a point. Using Proposition 6.2, $\tilde{f}: D^n \to \tilde{S}^n$ yields a map

$$\tilde{\tilde{f}}: S^n \to \tilde{S}^n$$

such that $p \cdot \tilde{\tilde{f}}: S^n \to S^n$ is the identity. (In fact, if $x \neq x_0$, $p \cdot \tilde{\tilde{f}}(x) = p \cdot \tilde{f}(x) = f(x)$, with x_0 being also the class of S^{n-1} in the quotient of D^n.) Then clearly $p \cdot \tilde{\tilde{f}}$ is the same map as the map, which $f: D^n \to S^n$ yields, via Proposition 6.2, on S^n, that is 1).

Now, apply the functor $\pi_1(\ )$ to the maps

$$S^n \xrightarrow{\tilde{\tilde{f}}} \tilde{S}^n \xrightarrow{p} S^n$$

whose composition is the identity. The homomorphism

$$\pi_1(S^n, x_0) \to \pi_1(\tilde{S}^n, \tilde{x}_0) \to \pi_1(S^n, x_0)$$

is the identity. But $\tilde{S}_n$ is the universal covering space of S^n, so $\pi_1(\tilde{S}^n, \tilde{x}_0) = 0$. We conclude at once

$$\pi_1(S^n, x_0) = 0.$$

Corollary 9.4

Let $x_0 \in P^2$, the projective plane (more generally P^n). Then

$$\pi_1(P^2, x_0) = Z_2,$$

the group of integers modulo 2.

Proof. $p: S^2 \to P^2$ (more generally $S^n \to P^n$) is a 2-sheeted covering space. S^2 is simply connected. By Corollary 9.1, the trivial subgroup has index 2 in $\pi_1(P^2, x_0)$. Hence $\pi_1(P^2, x_0)$ has 2 elements, and, of course, Z_2 is the only group with 2 elements.

Corollary 9.5

There is *no* $f: S^2 \to S^1$ which is a continuous map and has the property that $f(-x) = -f(x)$ for all $x \in S^2$, the unit sphere in R^3.

Proof. Suppose it were the case. Then f defines a

$$\tilde{f}: P^2 \to P^1 \equiv S^1$$

by $\tilde{f}(\{x\}) = \{f(x)\}$. I claim that a path which runs from $\tilde{x}_0$ to $-\tilde{x}_0$ in S^2 projects to a loop in P^2 which represents a non-zero element in $\pi_1(P^2, x_0)$. For if the loop is null homotopic, it will lift to a closed loop, by our earlier theorems, in contradiction to the uniqueness of path lifting. These remarks apply equally to S^1.

This implies that (assuming $f(-x) = -f(x)$) non-zero elements in $\pi_1(P^2, x_0)$ are mapped to non-zero elements in $\pi_1(S^1, y_0)$. In other terms, $\bar{f}_\#$ is 1-1.

But there is no 1-1 homeomorphism of Z_2 to Z. (If there were, Z would have a non-zero element n so that $n + n = 0$.)

(For $\pi_1(S^1, y_0) \simeq Z$, we refer to the next chapter, or to the "Problems" at the end of this chapter.)

Corollary 9.6 (Special case of Borsuk-Ulam Theorem)

Given any continuous map $g: S^2 \to R^2$, there is $x \in S^2$ so that

$$g(x) = g(-x).$$

Proof. Assume it is false. Set

$$\frac{g(x) - g(-x)}{\| g(x) - g(-x) \|} = f(x).$$

(Note that the denominator, which is the distance of $g(x) - g(-x)$ to $(0, 0)$, is positive for all $x \in S^2$.)

$$f(-x) = \frac{g(-x) - g(x)}{\| g(-x) - g(x) \|} = -\frac{g(x) - g(-x)}{\| g(x) - g(-x) \|} = -f(x).$$

Thus, the assumption that Corollary 9.6 is false would contradict Corollary 9.5.

Remarks. One can graphically describe this by saying that "there are two diametrically opposed points, on the surface of the earth, with the same temperature and pressure."

Problems

1. Show that the plane is a covering space of the torus, under the map which associates to each pair (x, y), the pair $(x - [x], y - [y])$ where $[x]$ means the largest integer less than or equal to x).
2. More generally, show that if $p_x: \tilde{X} \to X$ and $p_y: \tilde{Y} \to Y$ are covering spaces, then

$$p_x \times p_y: \tilde{X} \times \tilde{Y} \to X \times Y$$

   is also a covering space.
3. Suppose $A \subseteq X$ is an arcwise connected and locally arcwise connected subspace of X, and suppose $p: \tilde{X} \to X$ is covering space.
   Prove $p \mid p^{-1}(A): p^{-1}(A) \to A$ (in other words the part of the cover-

ing space lying over A) is a covering space, if and only if $p^{-1}(A)$ is arcwise connected.
4. Let X be an arcwise connected, locally arcwise connected space. Suppose $x_0 \in X$ and X is contractible to x_0. Prove that X has no covering space other than itself. (Use the relation between fundamental group and covering spaces. This is very easy).
5. Let X satisfy all the conditions of Theorem 9.2 and also be a topological group. Prove that the covering spaces constructed in that theorem are also topological groups. (*Hint:* Show how the assumption that X is a topological group gives rise to a multiplication on the paths which begin at $e \in X$. Prove that this yields a well defined multiplication on $\tilde{X}$).

The following long, but not difficult, exercise is an alternate computation that $\pi_1(S^1, x_0) \approx Z$. More complete computational methods are in the next chapter.

6. $\pi_1(S^1, x_0) \approx Z$. To establish this, prove the following pieces of a full analysis of this group (and more).
   a) The covering space $p \colon \mathbb{R} \to S^1$; $p(x) = e^{2\pi i x}$, is the universal covering space of S^1. If $\pi \colon \tilde{X} \to S^1$ is any covering space, there is a lift
   $$\Gamma \colon \mathbb{R} \to \tilde{X}$$
   so that $\pi \cdot \Gamma = p$. $\Gamma \colon \mathbb{R} \to \tilde{X}$ is also a covering space.
   b) From a), every covering space of S^1 is a 1-manifold homeomorphic either to S^1 or to $\mathbb{R}$.
   c) The only non-zero subgroups of $\pi_1(S^1, x_0)$ are isomorphic to $\pi_1(S^1, x_0)$. $\pi_1(S^1, x_0)$ is countable.
   d) S^1 is a topological group. $\pi_1(S^1, x_0)$ is Abelian.
   e) The number of generators of a group is the minimal number of elements (not e), say $x_1, \cdots, x_k$, so that products taken from $x_1, x_1^{-1}, x_2, x_2^{-1}, \cdots, x_k, x_k^{-1}$ give every element. Show that $\pi_1(S^1, x_0)$ must admit a single generator in this sense.
   f) Prove $\pi_1(S^1; x_0) \approx Z$. (Show Z is the only countable group with one generator. Show that c) would be violated, if $\pi_1(S^1, x_0)$ had more than one generator.)

CHAPTER 10

Calculation of Some Fundamental Groups. Applications

In our final chapter, we explore some computations and the methods which yield them. In particular we want to learn how to figure out $\pi_1(X, x_0)$ for a large class of spaces. The classical approach was to show how to find an algorithm for computing $\pi_1(X, x_0)$, when X was a nice space—for example triangulable. More recently, it was observed that most of the computations could be carried out effectively in a broader class of spaces, using the theorem of Seifert and van Kampen. This theorem is a classic for calculating the fundamental group of a union of two spaces, and it is known to hold for a much broader class of spaces than triangulable.

Unfortunately, the general theorems sacrificed some of the computational precision of the earlier methods—as a price for the increased generality. Nowadays, we are witnessing a revival in algorithmic methods of computation, undoubtedly brought on by the growth of machine computation. Once again, it is acceptable—in mathematical company—to take a small loss in generality in exchange for the possibility of a precise method of computation. In this spirit, I have decided to limit this chapter to the study of the fundamental group of simplicial complexes or—as they are sometimes called—triangulable spaces (our full definition is given below). Although we do not push this viewpoint to the limit, the reader may easily check that our Theorem 10.2 yields quickly a computer program for determining the generators and relations of $\pi_1(X, x_0)$, when X is a finite simplicial complex.

Various specific applications will follow at once from the computations. For example, we will be able to find all covering spaces of certain spaces. In addition, the fundamental group may be used as an invariant to deter-

Calculation of Fundamental Groups 181

mine when two spaces are not homeomorphic (or more generally not of the same homotopy-type, discussed in references, such as the book of Dold). This quickly implies that no two distinct surfaces, expressed with fewer than 3 cross-caps (see Proposition 7.6) can possibly be homeomorphic.

Definition 10.1

Let $X_1, \cdots, X_{n+1}$ be $(n + 1)$, distinct points in $\mathbb{R}^m$. That is each X_i is an m-tuple of real numbers. We note that the coordinate-wise difference of two such points defines a vector in $\mathbb{R}^m$.

We say that $X_1, \cdots, X_{n+1}$ are *independent*, if the n vectors

$$X_2 - X_1, X_3 - X_1, \cdots, X_{n+1} - X_1$$

are linearly independent vectors (see any basic algebra text). (Check that this is independent of the order involved.)

Definition 10.2

An n-simplex is the convex hull of $(n + 1)$ independent points in $\mathbb{R}^m$. In other words, it is the intersection of all closed convex sets which contain these points. We say that the $(n + 1)$-points span the simplex.

It is convenient to describe these simplexes in terms of some coordinates. The barycentric coordinates in the next definition and proposition are motivated by an analogy with elementary mechanics.

Proposition 10.1

Let $X_1, \cdots, X_{n+1}$ be independent points in $\mathbb{R}^m$. The set of points

$$\lambda_1 X_1 + \cdots + \lambda_{n+1} X_{n+1} \in X$$

(where multiplication of an n-tuple by λ means multiply each place by λ, and addition means coordinate-wise addition), subject to the conditions
 a) $\lambda_i \geq 0$ for each i
 b) $\sum_{i=1}^{n+1} \lambda_i = 1$
is precisely the n-simplex spanned by these points.

Proof. It is trivial to check that this set of points is closed. To verify that it is convex, consider

$$\lambda_1 X_1 + \cdots + \lambda_{n+1} X_{n+1}$$

and

$$\mu_1 X_1 + \cdots + \mu_{n+1} X_{n+1},$$

both belonging to this set. We calculate

$$t(\lambda_1 X_1 + \cdots + \lambda_{n+1} X_{n+1}) + (1-t)(\mu_1 X_1 + \cdots + \mu_{n+1} X_{n+1})$$
$$= (t\lambda_1 + (1-t)\mu_1) X_1 + \cdots + (t\lambda_{n+1} + (1-t)\mu_{n+1}) X_{n+1}.$$

Clearly each coefficient is positive; but note that

$$\sum_{i=1}^{n+1} (t\lambda_i + (1-t)\mu_i) = t \sum_{i=1}^{n+1} \lambda_i + (1-t) \sum_{i=1}^{n+1} \mu_i = t + (1-t) = 1,$$

proving convexity.

Our set being closed and convex, and trivially containing all the points $X_1, \cdots, X_{n+1}$, must contain the simplex spanned by these points. We must prove, on the other hand, that the simplex contains this set of points.

To this end, notice that if we ignore a point, say X_1, then the remaining n points span an $(n-1)$-simplex in the linear subspace of R^m which contains $X_2, \cdots, X_{n+1}$ and which is homeomorphic to R^{n-1}. Similarly, any smaller subcollection of points spans a suitable, lower dimensional simplex.

Now, let us rewrite (assuming $1 > \lambda_1 > 0$)

$$\lambda_1 X_1 + \cdots + \lambda_{n+1} X_{n+1} = \lambda_1 X_1 + (1-\lambda_1)$$
$$\cdot \left[\left(\frac{\lambda_2}{1-\lambda_1} \right) X_2 + \cdots + \left(\frac{\lambda_{n+1}}{1-\lambda_1} \right) X_{n+1} \right].$$

This displays our point as lying on the line from X_1 to the $(n-1)$-simplex spanned by $X_2, \cdots, X_{n+1}$; clearly, we may assume by induction that

$$\left(\frac{\lambda_2}{1-\lambda_1} X_2 + \cdots + \frac{\lambda_{n+1}}{1-\lambda_1} X_{n+1} \right)$$

lies on the $(n-1)$-simplex spanned by $X_2, \cdots, X_{n+1}$.

But then our conclusion is immediate because a simplex is defined as convex. (Check!) In other words, we have shown that a point of the form in question must also lie on the simplex. (Check that the assumption on λ_1 is in reality no restriction on the generality of the proposition here.)

Definition 10.3

The numbers $\lambda_1, \cdots, \lambda_{n+1}$ are the *barycentric coordinates* of the point in the simplex.

The set

$$\{\lambda_1 X_1 + \cdots + \lambda_{n+1} X_{n+1} \mid \lambda_j = 0\}$$

is called the *j-th face*. It lies opposite the j-th vertex X_j.

The *barycenter of a simplex* is the point

$$b = \frac{1}{n+1} X_1 + \cdots + \frac{1}{n+1} X_{n+1}.$$

Calculation of Fundamental Groups

Remark. a) These coordinates will become particularly useful when we divide simplexes into smaller simplexes by the process of barycentric subdivision. This is essential for our main approximation Theorem 10.1.

b) The barycenter is none other than the center of gravity of the mass configuration which assigns equally heavy particles to all of the vertices.

c) A simple example would be

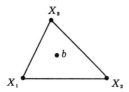

The barycentric coordinates are particularly useful in describing mappings. The basic illustration of that thought is the following proposition.

Proposition 10.2

Any two n-simplexes are homeomorphic.

Proof. Let σ^n and τ^n be the two n-simplexes, spanned by the points

$$X_1, \cdots, X_{n+1}$$

and

$$Y_1, \cdots, Y_{n+1}.$$

respectively.

Define $\phi: \sigma^n \to \tau^n$ by

$$\phi(\lambda_1 X_1 + \cdots + \lambda_{n+1} X_{n+1}) = \lambda_1 Y_1 + \cdots + \lambda_{n+1} Y_{n+1}.$$

ϕ is visibly continuous. It is trivial to construct an inverse by

$$\psi(\mu_1 Y_1 + \cdots + \mu_{n+1} Y_{n+1}) = \mu_1 X_1 + \cdots + \mu_{n+1} X_{n+1},$$

showing that the two are homeomorphic.

Note that these homeomorphisms are actually linear maps (take straight lines to straight lines).

We wish to glue simplexes together to build a "simplicial complex," of which a triangulable surface is a special case. We call faces, and faces of faces, etc., subsimplexes.

Definition 10.4

A *simplicial complex* K is a topological space which is a union of a finite number of subspaces, each of which is homeomorphic to some simplex, and which we write $\sigma_i^{n_i}$ (that is, the i-th simplex, having dimension n_i).

The simplexes of K, that is $\{\sigma_i^{n_i}\}$, are subject to two axioms.

a) If $\sigma_i^{n_i}$ is a simplex of K, and $\sigma_j^{n_i-1}$ is a face of $\sigma_i^{n_i}$ (see Definition 10.3), then $\sigma_j^{n_i-1}$ is also a simplex of K.

b) If $\sigma_i^{n_i}$ and $\sigma_k^{n_k}$ are simplexes of K, then either (i) $\sigma_i^{n_i} \cap \sigma_k^{n_k}$ is empty or (ii) $\sigma_i^{n_i} \cap \sigma_k^{n_k}$ is also a simplex of K, a subsimplex of $\sigma_i^{n_i}$ and $\sigma_k^{n_k}$.

Roughly put, a) says the list of simplexes of K is to be as complete as possible and b) says that two simplices can be glued together only by identifying together the points of a common subsimplex.

c) The topology on K is given by requiring that $C \subseteq K$ is closed, if and only if $C \cap \sigma_i^{n_i}$ is closed in $\sigma_i^{n_i}$ for each i.

Examples. 1) A graph—a finite union of closed intervals, glued together at their end points—is a simplicial complex. We call the *dimension of a complex* the maximal dimension of a simplex occurring in K. A graph is clearly a 1-dimensional simplicial complex.

2) A triangulable closed surface is obviously a 2-dimensional simplicial simplex.

3) If σ^n is an n-simplex, we define the boundary of σ^n to be the union of all the faces of σ^n, and write it $\partial \sigma^n$.

$\partial \sigma^n$ is an $(n-1)$-dimensional complex, homeomorphic to S^{n-1}. (Te check this, take the barycenter as the origin, and push radially from $\partial \sigma^n$ out to some larger sphere with the same center as the origin. Draw a picture in a suitable low dimensional case.)

Proposition 10.3

A simplicial complex K is homeomorphic to a subset of some Euclidean space $\mathbb{R}^m$.

Proof. Let $v_1, \cdots, v_m$ be the totality of the vertices of K, so that various subsets like $v_{i_1}, \cdots, v_{i_k}$ span simplexes of K. Of course, there may be some subsets which do not span a simplex belonging to K (for example, in $\partial \sigma^n$ of Example 3 above, all of the vertices do not span any simplex in the simplicial complex $\partial \sigma^n$).

Define a map ϕ from the set of vertices $v_1, \cdots, v_m$ to $\mathbb{R}^m$ by setting

$$\phi(v_j) = (0, \cdots, 1, \cdots, 0) \in \mathbb{R}^m$$

where 1 occurs in the j-th place. If x is a point of a simplex σ in K, where σ is spanned by $V_{i_1}, \cdots, V_{i_e+1}$, and where the barycentric coordinates of x in this simplex are given by

$$x = \lambda_1 V_i + \cdots + \lambda_{i_e+1} V_{i_e+1}$$

then we define

$$\phi(x) = \lambda_1 \phi(V_{i_1}) + \cdots + \lambda_{i_e+1} \phi(V_{i_e+1}).$$

(Check that this is well-defined and continuous.)

Calculation of Fundamental Groups

ϕ is clearly a homeomorphism on each simplex of K, because the points $\phi(V_1), \cdots, \phi(V_m)$ are clearly independent. But also note that the images of disjoint simplexes are disjoint. Hence ϕ is a 1-1 continuous map of K to $\mathbb{R}^m$; as K is visibly compact (it being a finite union of compact subspaces), we conclude that ϕ maps K homeomorphically onto $\phi(K)$.

Remarks. This shows that the finite simplicial complexes are subsets of Euclidean spaces. They are, in fact, very special subsets, and they enjoy some especially nice properties. For example, they are trivially *metric spaces;* it is not hard to see that they are locally connected, etc.

There is a procedure of making finer and finer complexes out of a given complex, which is called barycentric subdivision. This is particularly important for our basic "simplicial approximation theorem," Theorem 10.1. It constitutes a generalization of the obvious procedure of dividing a 1-simplex in half by adding a new vertex at the barycenter, and treating it as two simplexes, which are joined, end to end. In other words, we replace the complex.

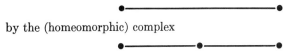

by the (homeomorphic) complex

To get at the definition, we need a bit of preliminary work.

Definition 10.5

Let σ^n be an n-simplex, K a subset of the boundary $\partial \sigma^n$ (which is defined as the union of all the faces). Let b be the barycenter.

We define the cone over K, with vertex b, to be the union of all closed straight line segments in σ^n, which begin at b and end at some point of K.

Example. The cone over K, with vertex b, is the shaded region

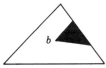

in the case where K is as follows

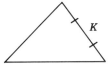

Definition 10.6

a) The barycentric subdivision of a zero simplex is to be itself.

b) The barycentric subdivision of a 1-simplex is given, as illustrated above, by two 1-simplexes, which are joined at the barycenter.

c) Suppose that the barycentric subdivision of any simplex of dimension less than n is defined. We define the barycentric subdivision of σ^n to be the following simplicial complex. The faces of σ^n are replaced by their barycentric subdivisions. If τ^j is a subsimplex of a face (so subdivided), we include the $(j+1)$-simplex formed by taking the cone over τ^j with vertex b. This is done for all such τ^j in the subdivided $\partial \sigma^n$.

It is clear (check!) that this is a simplicial complex which is homeomorphic to the original simplex.

Example. The barycentric subdivision of σ^2 is

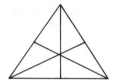

Definition 10.7

If K is a simplicial complex, the *barycentric subdivision of K*, written K^1, is the complex obtained by replacing each simplex of K by its barycentric subdivision (check that this is a complex).

Definition 10.8

The *mesh* of a complex is the maximum diameter of any simplex of K, where each simplex of K, being homeomorphic to a subset of a Euclidean space may be supposed to have that metric or distance. (Or by Proposition 10.3, we may suppose it to be a metric space lying in some $\mathbb{R}^m$ and take the definition with respect to that metric this way.)

Proposition 10.4

Let K be n-dimensional, i.e. the maximal dimension of a simplex of K is n.

$$\text{mesh of } K^1 \leq \frac{n}{n+1} (\text{mesh of } K).$$

Proof. It is easy to see that it suffices to do this for one simplex. The mesh of a simplex is trivially checked to be the maximum length of an edge. We will calculate the distance from the barycenter of a simplex to a vertex. (Check that this will suffice!)

Suppose the simplex is σ^n spanned by $X_1, \cdots, X_{n+1}$. Suppose also that

the vertex is
$$1 \cdot X_1 + 0 \cdot X_2 \cdots + 0 X_{n+1}$$
while the barycenter is
$$\frac{1}{n+1} X_1 + \cdots + \frac{1}{n+1} X_{n+1}.$$
The two points may be expressed as the following vectors (see Definition 10.1 and Definition 10.2)
$$1 \cdot (X_1 - 0) + 0 \cdot (X_2 - X_1) + \cdots + 0 \cdot (X_{n+1} - X_1)$$
and
$$1 \cdot (X_1 - 0) + \frac{1}{n+1} (X_2 - X_1) + \cdots + \frac{1}{n+1} (X_{n+1} - X_1)$$
(the parentheses being vectors).

The difference of the two vectors, subtracting the first from the second, is
$$\frac{1}{n+1} (X_2 - X_1) + \cdots + \frac{1}{n+1} (X_{n+1} - X_1).$$
By the usual triangle inequality, this is, in length, less than
$$\frac{n}{n+1} \max_{2 \leq i \leq n+1} |X_i - X_1|.$$
But $|X_i - X_1|$ is smaller than or equal to the mesh of the simplex σ^n, spanned by these vertices.

This completes the proof.

Corollary 10.1

Define $K^{(m)} = (K^{(m-1)})^1$ inductively as the m-fold iterated barycentric subdivision of K. Then the mesh of $K^{(m)}$ approaches 0 as m goes to infinity.

Proof. Note that the operation of barycentric subdivision does not change the dimension of a simplicial complex. The proof follows at once because
$$\lim_{m \to \infty} \left(\frac{n}{n+1} \right)^m = 0.$$
We are now in a position to head toward the basic approximation theorem (Theorem 10.1), which approximates continuous maps between simplicial complexes by maps which are linear on every simplex (frequently called *simplicial maps* or *piecewise-linear maps*). To build this approximation, we need to study a nice cover of a simplicial complex by open sets, which are related to the simplicial structure. This is achieved by the notion of a *star*.

Definition 10.9

Let K be a simplicial complex, $x_0 \in K$ a vertex. $St(x_0)$ is defined to be the complement of

$$\bigcup_{\substack{\tau \in K \\ x_0 \cap \tau = \phi}} \tau$$

that is to say, the complement of the subset of K which is obtained by taking the union of all simplexes which do not meet x_0. $St(x_0)$ is called the *star* of x_0; it is the complement of a compact subset of K and is clearly then open.

Remarks. 1) If the complex K is a subdivided 1-simplex

```
•————————•————————•
a   σ¹    b   τ¹    c
```

with σ^1 and τ^1 being the two (closed) 1-simplexes, then

$$St(a) = K - \tau^1$$
$$St(c) = K - \sigma^1$$
$$St(b) = K - \{a\} - \{c\}$$

2) If K is the boundary of a tetrahedron,

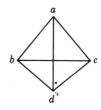

with vertexes a, b, c, d, then $St(a)$ is the complement of the closed simplex which is spanned by b, c, and d.

3) It is clear that the collection of the stars of every vertex of K covers K.

The following proposition is a preliminary version of the simplicial approximation theorem.

Proposition 10.5

Let K and L be simplicial complexes and $f: K \to L$ be a continuous map. Suppose that for every vertex x in K, there is a vertex y in L such that

$$f(St(x)) \subseteq St(y).$$

In other words, f sends stars into (subsets of) stars.

Then there is a map $g: K \to L$ such that
1) g maps vertices of K to vertices of L. If a point p, with barycentric expression

$$\lambda_1 x_1 + \cdots + \lambda_{k+1} x_{k+1},$$

lies on the simplex spanned by $x_1, \cdots, x_{k+1}$ of K, then

$$g(p) = \lambda_1 g(x_1) + \cdots + \lambda_{k+1} g(x_{k+1})$$

on a simplex which is spanned by the vertices $g(x_1), \cdots, g(x_{k+1}) \in L$.

2) For any point $x \in K, f(x)$ and $g(x)$ belong to a (single) star of some vertex of L.

Proof. For each vertex $x_i \in K$, select $y_i \in L$ with $f(St(x_i)) \subseteq St(y_i)$. Define

$$g(x_i) = y_i.$$

Let p now be any other point in K. Select a maximal simplex σ of K, with p in the interior of σ. Then, if the vertices of σ are say $x_1, \cdots, x_{k+1}$,

$$p \in St(x_1) \cap \cdots \cap St(x_{k+1}).$$

(Check this!)

But then

$$f(p) \in f(St(x_1)) \cap \cdots \cap f(St(x_{k+1})) \subseteq St(y_1) \cap \cdots \cap St(y_{k+1}).$$

I claim that $y_1, \cdots, y_{k+1}$ must then be the vertices of a simplex of L. Whenever their stars have non-empty intersection assume by induction that $y_1, \cdots, y_k$ are the vertices of a simplex. Consider the simplex containing the vertices $y_1, \cdots, y_k$ and suppose y_{k+1} is not a vertex of any simplex which contains $y_1, \cdots, y_k$. It is easy to see that every simplex containing y_{k+1} must fail to meet some $St(y_i), 1 \leq i \leq k$. Considering the definition of $St(y_{k+1})$ it is clear that it doesn't meet

$$St(y_1) \cap \cdots \cap St(y_k)$$

(which is clearly made up of points which lie on simplexes that contain $y_1, \cdots, y_k$). Hence, we conclude that y_{k+1} is a vertex of a simplex which contains $y_1, \cdots, y_k$. But then $y_1, \cdots, y_{k+1}$ span a subsimplex of such a simplex of L, proving the claim. (Check!)

Getting back to our proof, we now know that $y_1 = g(x_1), \cdots, y_{k+1} = g(x_{k+1})$ span a simplex of L. Working with barycentric coordinates, we put

$$g(p) = \lambda_1 \cdot g(x_1) + \cdots + \lambda_{k+1} g(x_{k+1}),$$

whenever $p = \lambda_1 x_1 + \cdots + \lambda_{k+1} x_{k+1}$.

It is trivial that g is well-defined, continuous, and satisfies our condition 1.

Notice also that

$$g(p) \in St(g(x_1)) \cap \cdots \cap St(g(x_{k+1}))$$

so that $f(p)$ and $g(p)$ must belong to some common star. This completes 2) and the proof.

Remarks. 1) A map g, which satisfies 1) of this proposition, is called *simplicial*.

2) f and g approximate one another in that they never move further apart than the maximum radius of a star in L, which is the mesh. We now must interject a point-set theoretic lemma.

Lemma

Let X be a compact metric space, with $\{O_i\}$ a cover of X. Then there is a $\delta > 0$ (called the Lebesgue number) such that whenever $x \in X$, the open ball

$$B_\delta(x) = \{y \mid y \in X, d(x,y) < \delta\}$$

belongs entirely to some O_i.

Proof. Suppose not, i.e. for each $1/n$, there is x_n, with $B_{1/n}(x_n)$ not entirely in any O_i. Let x be an accumulation point of the x_n. I claim then that x belongs to no O_i, which will be a contradiction and give thus a proof of the lemma.

If it were the case that $x \in O_i$, choose $\epsilon > 0$ so that

$$x \in B_\epsilon(x) \subseteq O_i.$$

Choose x_n with $d(x, x_n) < \epsilon/2$ and $1/n < \epsilon/2$. Then

$$B_{1/n}(x_n) \subseteq B_{\epsilon/2}(x_n) \subseteq B_\epsilon(x) \subseteq O_i,$$

contradicting the presumption that no $B_{1/n}(x_n)$ was entirely in any O_i.

The lemma and the Proposition 10.5 will now yield our first main theorem.

Theorem 10.1 (Simplicial Approximation Theorem)

Let K and L be simplicial complexes (which we may assume are metric spaces as above), and let $f\colon K \to L$ be a continuous map.

Then there is a simplicial map, from the n-th barycentric subdivision $K^{(n)}$ to L,

$$S\colon K^{(n)} \to L,$$

i.e. S sends vertices to vertices and is linear on every simplex (as in (1) of Proposition 10.5), such that

1) For any $x \in K$, $S(x)$ and $f(x)$ lie in a single simplex of L.

2) S and f are homotopic.

Proof. Using the previous lemma, choose $\epsilon > 0$ such that every open ball $B_\epsilon(y)$ lies entirely within some open star of L. Choose $\delta > 0$ so that
$$f(B_\delta(x)) \subseteq B_\epsilon(f(x))$$
(because f is a continuous function and K is compact).

Using Corollary 10.1, choose n so large that every star of x in $K^{(n)}$ lies entirely within $B_\delta(x)$ for the above fixed δ. Regarding $K^{(n)}$ as a homeomorphic copy of K, our map f
$$f\colon K^{(n)} \to L,$$
will satisfy the condition
$$f(St(x_i)) \subseteq f(B_\delta(x_i)) \subseteq B_\epsilon(f(x_i)) \subseteq St(y_i)$$
for each $x_i \in K^{(n)}$ and suitable $y_i \in L$.

By Proposition 10.5, we may find a simplicial S which satisfies condition 1). We must now establish 2), to complete the proof.

But if we look at the proof of Proposition 10.5, we see that $S(x)$ and $f(x)$ lie on a common simplex of L, for each x, because they lie on the stars of the y_i corresponding to x_i which span the simplex which contains x. Regardless of the fact that these two points may lie simultaneously on various simplexes, there is a unique straight line in L from $S(x)$ to $f(x)$. We call this line $\ell(x)(t)$, expressed as a map (linear) of I to L. That is
$$\ell(x)(0) = S(x)$$
and
$$\ell(x)(1) = f(x).$$
Define a homotopy
$$F\colon K^{(n)} \times I \to L$$
by
$$F(x, t) = \ell(x)(t).$$
Then easily, $F(x, 0) = S(x)$ and $F(x, 1) = f(x)$, finishing the proof.

Remarks. 1) We may assume, checking over the proof of this theorem, that if f maps a vertex $x \in K$ to a vertex $y \in L$, then $S(x) = y$. For we need only send x to a vertex, whose star contains the image of the star of x, and y does have this property.

2) If dim K means the maximum dimension of a simplex which occurs in K, and $f\colon K \to L$ with dim $K <$ dim L, then f is homotopic to a map which is not onto. In fact, such a simplicial S, from Theorem 10.1 is not onto, under these circumstances.

192 Topology

3) We now may give a new proof that $\pi_1(S^n, x_0) = \{e\}$ for $n > 1$, which was first proved at the end of the last chapter. Our Remarks 1 and 2 above show that, if $\alpha: I \to S^n$ represents a class in $\pi_1(S^n, x_0)$, α is homotopic (respecting base point) to a simplicial map S. Since dim I is 1, which is less than dim $S^n = n$, the range of S is not all of S^n. Suppose $Z_0 \in S^n$ is not in the image of S. Then S factors through $S^n - Z_0$, that is

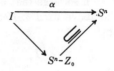

Since $S^n - Z_0$ is homeomorphic to $\mathbb{R}^n$, and therefore contractible to x_0, we see at once that every such map α is null-homotopic. Then $\pi_1(S^n, x_0)$ is trivial.

4) Let the complex $A_1 \subseteq K$ be the union of all 0-dimensional and 1-dimensional simplexes of K. (This is sometimes referred to as the 1-skeleton of K.) Then if the inclusion map $A_1 \subseteq K$ is denoted i, we have that

$$i_\#: \pi_1(A_1, x_0) \to \pi_1(K, x_0)$$

is onto. This follows at once from the Theorem 10.1, because that shows that if $\{\alpha\} \in \pi_1(K, x_0)$, α is homotopic to a simplicial map, which sends vertices to vertices and 1-simplexes to 1-simplexes, or in other words, maps I into the 1-skeleton of K, namely A_1.

5) Let A_2 be the union of all simplexes, of dimension less than or equal to 2, in a connected complex K. It will follow, from our next theorem, that the inclusion map $i: A_2 \to K$ induces an isomorphism

$$i_\#: \pi_1(A_2, x_0) \xrightarrow{\approx} \pi_1(K, x_0).$$

Our next theorem will show that the study of the fundamental group of a simplicial complex may be reduced to a study of simplicial paths. This will yield a computational method of getting at the fundamental group of a simplicial complex.

Definition 10.10

Let K be an arcwise connected simplicial complex with base point x_0. A *simplicial loop in K* is a simplicial map from some subdivision of the unit interval I, which is a 1-simplex, to K, which begins and ends at x_0. In other terms, a simplicial loop may be regarded as a continuous map from the interval $[0, n]$ regarded as a union of 1-simplexes such as $[0, 1], [1, 2], \cdots$, $[n - 1, n]$ to K, which maps each subsimplex such as $[i - 1, i]$ linearly onto a 1-simplex of K, and which begins and ends at the base point. (Recall that I and $[0, n]$ are homeomorphic. Of course, x_0 is a vertex.)

Denote the set of all such simplicial loops by $\Omega_S(K, x_0)$.

We wish to define an equivalence operation on $\Omega_S(K, x_0)$, which makes homotopic maps equivalent, yet uses the simplicial structure of K.

Definition 10.11

We denote simplicial loops as successions of vertices, where each two consecutive vertices are connected by a 1-simplex which is an image of some unit interval. For example, $a_1 a_2 \cdots a_1$ means the 1-simplex a_1 to a_2, followed by a_2 to a_3, etc. (Draw a picture!)

We define three elementary equivalence operations.

i) If a vertex occurs twice, it may be written only once. For example $a_1 a_2 a_2 a_3 \cdots$ is equivalent to $a_1 a_2 a_3 \cdots$.

ii) In any word, the letters $a_i a_{i+1} a_i$ may be replaced by just a_i.

iii) If a_i, a_{i+1}, and a_{i+2} are the vertices of a 2-simplex in K, then the letters $a_i a_{i+1} a_{i+2}$ may be replaced by $a_i a_{i+2}$.

Two simplicial loops are equivalent, if one may be obtained from another by a succession of these elementary operations or their inverses. (Check that this is an equivalence relation!)

The quotient set of $\Omega_S(K, x_0)$ by this equivalence relation is trivially checked to be a group under composition of paths. It is the *edge path group*, which we denote as

$$E_1(K, x_0).$$

It is clearly a functor on the category of connected simplicial complexes with base point (a vertex), and simplicial maps which preserve base point, to the category of groups. The trivial path at the base point acts an identity.

Examples. 1) If $a_3 a_4 a_5$ are the vertices of a 2-simplex in K, then $a_1 a_2 a_2 a_3 a_4 a_5 a_6 a_7 a_6 a_1$ is equivalent to $a_1 a_2 a_3 a_5 a_6 a_1$, as simplicial loops in K.

2) Consider S^1 as the simplicial complex abc which is the boundary of a 2-simplex (which itself is not in the complex). One easily sees that any element in $\Omega_S(S^1, a)$ is equivalent to a simplicial loop which wraps around S^1 a certain number of times (either clockwise or counterclockwise).

By sending such a loop to the number $+n$, in case it wraps around n times counterclockwise, or $-n$ if clockwise, we construct an isomorphism of $E_1(S^1, a)$ with the group of integers.

3) In the Möbius band M, which we write as a complex

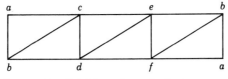

with base point b, any simplicial loop is equivalent to one which uses only the simplexes bd, de, and eb. (Check!) (Of course, each of these simplexes may be used more than once.)

Our main result is

Theorem 10.2

Let K be a connected simplicial complex with a vertex x_0 as base point. Then we have an isomorphism of groups

$$E_1(K, x_0) \approx \pi_1(K, x_0).$$

Proof. Let $x_0 a_1 a_2 \cdots x_0$ be a simplicial loop in K, which is in fact a map $S \colon [0, n] \to K$. Let $\ell \colon I \to [0, n]$ be the unique, linear, order-preserving homeomorphism, given by $x \to n \cdot x$. Define

$$\phi \colon \Omega_S(K, x_0) \to \Omega(K, x_0)$$

by

$$\phi(S) = S \cdot \ell.$$

It is clear that equivalent simplicial loops are homotopic, because each elementary equivalence operation of Definition 10.11 takes such a loop into a homotopic one. Equally clear is the fact that the map

$$\phi_\# \colon E_1(K, x_0) \to \pi_1(K, x_0),$$

which is defined by $\phi_\#(\{S\}) = \{\phi(S)\}$ is a homomorphism. (Check!)

Given any $\{\alpha\} \in \pi_1(K, x_0)$, the simplicial approximation theorem (Theorem 10.1) yields a simplicial map $\alpha' \colon I^{(n)} \to K$ which is homotopic to α with end points left fixed, for α sends the end points to x_0, the base point. (Check!) But by reparametrizing, we see that α' is the image of a simplicial loop, which has as many segments as there are pieces of the simplicial map α'. (Recall $I^{(n)}$ is the n-fold barycentric subdivision of I.) This shows at once that $\phi_\#$ is onto.

The hard work now consists in proving that $\phi_\#$ is 1-1. This is the same as showing that if two simplicial loops are homotopic, with end points fixed, both regarded as maps of I, then they were actually equivalent in the precise sense of Definition 10.11.

First, we note that the two simplicial loops—which we assume are homotopic—may in fact be regarded as having the same number of segments. This follows because we may invoke the elementary operation (ii) from Definition 10.11 to increase the number of segments, without changing the equivalence class of the loop.

Second, observe that if we start with a simplicial loop, and replace it by a simplicial approximation to itself by using the simplicial approximation theorem (Theorem 10.1) on a complex, which contains the interval (which is the domain of the loop) as a subcomplex, then the new loop may be assumed to be equivalent to the old in the sense of Definition 10.11. This follows from the fact that the procedure of Theorem 10.1 allows us to choose for each $x \in K^{(n)}$ any vertex $y \in L$ with $f(St(x)) \subseteq St(y)$ so that were $f(x)$ already a vertex, we may choose $y = f(x)$. Also, one easily checks that the resulting simplicial map is equivalent (in sense of Definition 10.11) to the old.

Thirdly, note that $I \times I$ is obviously a complex (see the pictures below).

To complete the proof, let $F: I \times I \to K$ be a homotopy of our two simplical loops, and let $\tilde{F}: (I \times I)^{(n)} \to K$ be a simplicial approximation. Our remarks above show that we may assume that the map $\tilde{F}$, for $u = 0$ and for $u = 1$ equivalent to our original homotopic simplicial loops. Consider $(I \times I)^{(n)}$, which we may clearly assume to have the form

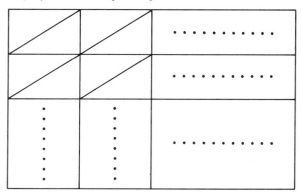

by observing that all that is required, of iterated subdivision, is that the mesh goes to 0. Take two paths which are simplicial, which begin at the left ($t = 0$) and end at the right ($t = 1$), and which only differ at one simplex. For example

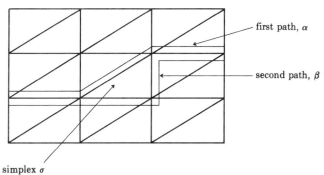

where the two paths only differ on the simplex σ. Call them α and β.

196   Topology

I claim that $\tilde{F} \cdot \alpha$ and $\tilde{F} \cdot \beta$ are equivalent in the sense of Definition 10.11. In fact, since $\tilde{F}$ is a simplicial map $\tilde{F}(\sigma)$ is a simplex of K. If it is a 2-simplex, $\tilde{F} \cdot \alpha$ and $\tilde{F} \cdot \beta$ will be equivalent in the sense of (iii). If it is a 0-simplex (all vertices mapped to a point) or a 1-simplex (two vertices to a point, one to another point), the loops are still clearly equivalent in the sense of Definition 10:11. (Check!)

But now, the loops in question are the compositions of $\tilde{F}$ with the paths which run across the bottom or the top. It is obvious that the bottom path may be brought to the top path through a succession of paths, each two consecutive ones being the same except for one simplex, as directly above. For example, writing the paths as ≫≫≫≫≫≫≫≫,

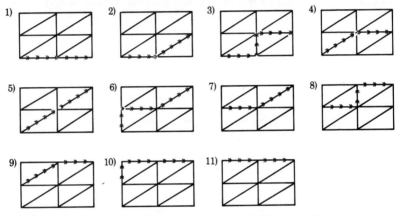

Note that the equivalence, between the 6th and 7th stages, is simply a use of Definition 10.11, (i), as the left and right vertical segments are assumed to be sent to the base point.

Since we have a succession of loops from one of our loops to the other, and since any two successive ones (which differ only on a simplex) are equivalent as in Definition 10.11, the theorem is complete.

These corollaries all use the nature of $E_1(K, x_0)$.

Corollary 1

Let K be an arcwise connected complex, with a vertex x_0 as base point. Then $\pi_1(K, x_0)$ is finitely generated. (That is, there are finitely many elements and inverses $u_1, u_1^{-1}, \cdots, u_k, u_k^{-1}$ in $\pi_1(K, x_0)$ such that any element is a product of these.)

Proof. There are finitely many simplicial loops which consist of a path α from the base point to a_1, with no simplex traversed more than once, followed by a simplicial loop $a_1 a_2 a_3 a_1$, followed by α^{-1}. (Recall that K has finitely many simplexes.) Denote these $\ell_1, \cdots, \ell_k$. Given an arbitrary simplicial loop ℓ, we proceed along ℓ as long as it is 1-1. Then we replace ℓ by $\alpha\, a_1 a_2 a_3 a_1\, \alpha^{-1} \alpha$ and continue on ℓ (α being a 1-1 map). This is illustrated as follows. We consider a 1-complex

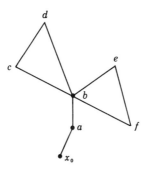

The loop $x_0 abcdbefbax_0$ would be replaced by $(x_0 abcdbax_0)(abefbax_0)$ which is the simplicial loop $x_0 abcdbax_0$ of our special type, followed by $abefbax_0$. Note that the second loop has fewer segments.

In other words, whenever the loop crosses back on itself, we replace it by some ℓ_i followed by a shorter loop. Continuing in this way, any loop is equivalent to a succession of the ℓ_i's. Hence, the corollary is established.

Corollary 2

Let A_2 be the union if all 0, 1, or 2-dimensional simplexes of K. Let $i: A_2 \to K$ be the inclusion map. Then

$$i_\#: \pi_1(A_2, x_0) \to \pi_1(K, x_0)$$

is an isomorphism.

Proof. By Remark 4 following Theorem 10.1, we get that i must be onto, as this is even the case without the 2-dimension simplexes. But the equivalence relation of Definition 10.11, for simplicial loops, only uses the simplexes of K whose dimension is less than or equal to 2. Thus, the map

$$i_\#^s: E_1(A_2, x_0) \to E_1(K, x_0)$$

is 1-1. In view of Theorem 10.2, the map in question $i_\#$ is also 1-1.

Corollary 3

$\pi_1(S^1, x_0)$ is isomorphic to the integers.

Proof. Direct application. The integer $+n$ corresponds to the counterclockwise path n times around.

Corollary 4

Let K be a connected simplicial complex, with base point the vertex x_0. Let A and B be connected subcomplexes (unions of some simplexes, and their subsimplexes) such that

$$A \cap B = x_0$$
$$A \cup B = K.$$

Then $\pi_1(K, x_0)$ is the free product of the groups $\pi_1(A, x_0)$ and $\pi_1(B, x_0)$. (The free product of two groups consists of the group made up of words from elements in each group (and their inverses). These words are equivalent precisely when one may be obtained from another by a finite number of relations, each of which involves the elements of only one group at a time.)

Proof. Consider an element of $E_1(K, x_0)$. It is represented by a simplicial loop, which obviously decomposes into a composition of smaller simplicial loops with each lying entirely within A or entirely within B. The corollary is then immediate. (As a reference to material on free products, I suggest Kurosh's book (see Bibliography).)

Corollary 5

If A, B and K are as in Corollary 3, and neither A nor B is simply connected, that is both have non-trivial fundamental groups, then $\pi_1(K, x_0)$ is non-commutative.

Proof. If $u \in \pi_1(A, x_0)$ and $v \in \pi_1(B, x_0)$ are non-identity elements, then $u \cdot v$ and $v \cdot u$ are always different in the free product.

As a special case, $S^1 \vee S^1$ (the union of two circles with a point in common) has a noncommutative fundamental group (compare with Problem 2 at the end of Chapter 8). We refer to $S^1 \vee S^1$ loosely as a "figure eight."

Corollary 6

The fundamental group of a handle (torus less an open disc) is a free group on two generators (free product of the integers with itself).

Proof. Use our theorem on the complex

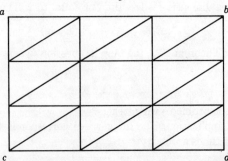

with the open middle square missing, and ab identified with cd and ac identified with bd. Connect the closed paths ac and ab to the base point to get the generators.

Corollary 7

The fundamental group of the Möbius band is isomorphic to the integers. Depending on the choice of isomorphism and choice of direction, the path around the boundary represents ± 2.

Proof. Direct use of our Theorem 10.2.

Corollary 8

Let K be the union of two connected subcomplexes. A and B such that $A \cap B$ is a circle S^1, which contains the base point. Then $\pi_1(K, x_0)$ is the quotient of the free product of $\pi_1(A, x_0)$ and $\pi_1(B, x_0)$ by the normal subgroup generated by the elements

$$i_\#^A(-n) \cdot i_\#^B(n)$$

where $n \in \pi_1(S^1, x_0)$ and $i^A : A \cap B \to A$ is the inclusion (and similarly i^B).

Proof. As in proof by Corollary 1, express the fundamental groups in terms of a loop around S^1 and then other loops which lie either in A entirely or in B entirely. Note that if these loops are chosen to be simplicial, the only relations of these generators are either relations from A, or relations from B, or the relations, which take an element in the fundamental group of S^1, regarded as belonging to A, and regard it as an element of the fundamental group of B; as S^1 is assumed to have the same orientations according as whether it is regarded in A or in B, these relations identify $i_\#^A(+n)$ with $i_\#^B(n)$.

Clearly, a group with these generators and relations is obtained from the free product of $\pi_1(A, x_0)$ and $\pi_1(B, x_0)$ by adding the new relations

$$i_\#^A(-n) \cdot i_\#^B(n) = e$$

proving the corollary. (For generators and relations, consult the text of A. Kurosh.)

Remarks. a) Corollaries 6, 7 and 8 yield directly a method for describing the fundamental group of any compact triangulated surface. Details are left to the problems below (which actually suggest a different method utilizing Theorem 10.2 directly).

b) These methods can be used to give a alternate proof of the fact (used in Theorem 7.1) that if a triangulated 2-manifold is a quotient of a disc, then the identification on the boundary consists of identifying *pairs* of edges. (At that time, this had to be done by arguing that if more than two edges were identified, the result could not be homeomorphic to a neighborhood in the plane, for if a suitable line segment were removed, the complement would be connected in one case but not in the other.)

This comes down to showing that a fin obtained by gluing more than two triangles along an edge is *not* homeomorphic to a fin obtained by gluing together two triangles. For example

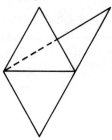

is *not* homeomorphic to

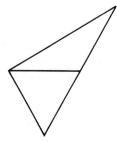

To do this, we choose a central point y_0 of such a fin K, and calculate $\pi_1(K - y_0, x_0)$, x_0 lying on the boundary.

Clearly $K - y_0$ has the homotopy-type of the boundary, that is there are maps both ways, whose compositions are homotopic to the respective identities. For example

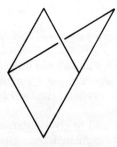

where the interiors of the triangles and the horizontal line have been removed. This has the homotopy-type of 2 circles joined at a point. (Check!)

In general, if K is a fin with n-triangles, $K - y_0$ has the homotopy-type of $(n - 1)$ circles attached at a point. Naturally, if such spaces have the same homotopy-type, they have isomorphic fundamental groups. (Check!)

The fact that if $n > m$, the corresponding fins, with y_0 deleted are not homeomorphic, is immediate, because $(n - 1)$ circles glued at a point and fewer than $(n - 1)$-circles glued at a point have different fundamental groups (use Corollary 4).

c) Students familiar with group theory will see immediately that Corollary 1 may be strengthened to show that $\pi_1(K, x_0)$ is finitely presented.

d) Corollary 8 may be strengthened easily to cover the case where $A \cap B$ is any connected subcomplex of K. (Consult the book of W. S. Massey.)

Problems

1. The projective plane P^2 is a simplicial complex

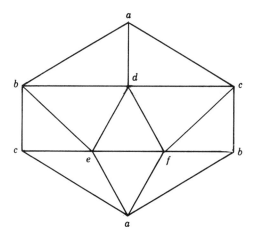

where the interiors of the triangles are included and the appropriate edges are to be identified. Use Theorem 10.2 to give a proof of the result (Corollary 9.4) that $\pi_1(P^2, x_0) \approx Z_2$, the group of integers modulo 2.

2. Using Theorem 8.2, show that the figure eight is not a topological group.

3. Calculate the fundamental group of a Klein bottle. Deduce that it is not a topological group.

4. Using the structure of 2-manifolds, show that the fundamental group of a sphere, with n handles, is generated by $2n$ elements $\alpha_1, \beta_1, \cdots, \alpha_n, \beta_n$, and the relation

$$\alpha_1\beta_1\alpha_1^{-1}\beta_1^{-1} \cdots \alpha_n\beta_n\alpha_n^{-1}\beta_n^{-1} = e.$$

(*Hint:* Apply Theorem 10.2 to

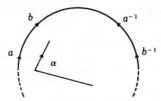

choosing for α the path indicated in the figure, etc. Note that the composite of all paths around the boundary is equivalent to the trivial path at the base point.)

5. Calculate the fundamental group of a sphere with n cross-caps. Observe that the distinct 2-manifolds described in Theorem 7.2, taking Proposition 7.6 into account, are all pairwise *not* homeomorphic.
6. Strengthen Corollary 4 to show that if a complex K is a union of connected subcomplexes A and B, with $x_0 \in A \cap B$, and $\pi_1(A \cap B) = \{e\}$, then $\pi_1(K)$ is the free product of $\pi_1(A)$ and $\pi_1(B)$.
7. Use Corollary 8 and Corollary 3 to show that $\pi_1(S^2, x_0) = \{e\}$.
8. Use Problem 7 and Corollary 8 to show that $\pi_1(S^n, x_0) = \{e\}$ for all $n > 1$. (Compare Corollary 9.3.)
9. Find all covering spaces of a Cartesian product of two projective planes.

Epilogue

To close this text, I would like to offer some hints and suggestions to the reader along a variety of lines. Throughout the book, I have stuck to a narrow scope in order to get across a certain basic curriculum. While I think one can make a very strong case for this choice of topics, at this level, there are many points which we have omitted. My goal here is to suggest how one can find these omitted topics and then to point to further study at a more advanced level. Keeping to an informal style, I will separate various topics into sections and refer to books by their author's names.

Foundations

Nowadays, this consists of logic, set theory, and category theory. A serious student of mathematics should see something of all three. Usually, logic and set theory appear together in one or more courses in a given curriculum. Both are very technical and can easily eat up a life-time of study. Most mathematicians—whose interest lies outside logic—would at least want to see some of this machinery in their student days, even though they rarely use it.

Category theory was originally set up to formalize some of the structure in topology (particularly algebraic topology). Somewhat later, F. Lawverre found a way to relate category theory to the foundations of mathematics, describing sets in these terms. This is elegant and interesting, but has not as yet displaced the traditional set theory. For texts on categories, I find B. Mitchell and the 2 volumes of H. Schubert (see Bibliography for details) helpful.

One word of warning: people who work in foundations have occasionally made inflated claims about the key role that their field plays for all mathematics, and even that they have systematized the way in which all mathematicians think. Of course, there is much in common in the thought of mathematicians, scientists, and so forth. But this writer (who personally met many mathematicians from the most distinguished professors to beginning students) has observed that most practicing mathematicians give little or no thought to the foundations of the field, once they begin their research work. While it may be debated as to whether this is healthy, logic and foundations do not at present figure in the everyday thoughts of most mathematicians.

Point-Set Topology

We have developed the main lines of thought. For elaborations in many directions, one could consult the books of Bourbaki, Kelley, etc. Here are some specific areas.

a) *Local properties*. A nice elementary treatment is available in the book of Hocking and Young; it contains a good description of some pathological phenomena.

b) For the many variants on *compactness*, one may find a broad discussion in the text of Dugundji. *Paracompactness* is most important, after compactness and local compactness.

c) For *dimension theory* and other interesting topics involving separable metric spaces, the book of Hurewicz and Wallman is fine (though a bit out of date).

d) There is a powerful and important school of American mathematicians, often from the South, who have studied an area of point-set topology which is frequently interesting even for subsets of the plane. See the American Mathematics Society colloquium volume of R. L. Moore. This frequently deals with unusual local properties of sets.

Most mathematicians, who work in other fields, would *only* consult these latter specialized areas when a specific problem arose.

Manifolds

This topic is a dominant force in much of the world's mathematics, in this half of the century. The topic is best exposed in connection with the study of *differentiable manifolds*. I recommend any text by J. Milnor (such as his book on Morse theory). The book of H. Whitney has much interesting material, from somewhat special points of view. Many books on differential geometry have useful information here; in fact, that subject constitutes an analytic study of manifolds endowed with special metrics.

Naturally, there is overlap with *topological groups*. *Lie groups* are the common ground between such groups and manifolds. The text of L. Pontryagin is perhaps the most clear and readable presentation, in an area where a great many mathematicians have tried their expository talents. For the overlap with point-set topology, the text of D. Montgomery and L. Zippin is excellent.

The natural generalization of the fundamental group is *homotopy theory*, which is a broad subject that studies an entire family of groups, which has the fundamental group as its first member. This field is an important part of algebraic topology, which I discuss subsequently. The natural generalization of covering spaces—*fibre spaces* or *fibre bundles*—has become an important area in mathematics. We have actually touched on some examples of fibre spaces—without mentioning them—in our text. A fibre space, just like a covering space, is locally like a product, but not necessarily globally a product. A covering space is a case where the fibre—the inverse image of a point—is a discrete space. A fibre bundle has some additional structure involving the action of a group; a simple example would be the Klein bottle.

Fibre bundles are useful in differential geometry, topological dynamics, Lie groups, algebraic topology, etc. They are now treated in many places; but in my opinion, the book of N. E. Steenrod remains the best thought-out exposition. The beginnings of this subject are completely accessible to someone who knows this book; the advanced parts, however, use considerable algebraic topology.

Algebraic Topology

This loosely means those areas where algebraic machinery—such as the fundamental group—is used to solve geometric problems. There are many good books now. For a general introduction to homology theory, algebraic topology of manifolds, etc., I recommend the book of A. Dold. For the special and advanced topic of homotopy theory, I would suggest the book of Hu. Spanier's text is a good encyclopedia.

For a remarkably elementary presentation of an advanced topic, namely K-theory, you could consult M. Atiyah's book (see Bibliography).

Roughly speaking, the subject studies various functors, such as homology and cohomology or K-theory, which are especially useful for solving specific geometric problems.

Finally, as in off shoot of algebraic topology, a new breed of algebra has come forth in recent years. *Homological algebra* was originally conceived to give a precise formulation to some of the methods developed in algebraic topology. It has grown considerably and must surely be taken as a field in its own right. An excellent text is the book of S. MacLane.

As a final and *very personal* attempt to suggest further lines of study, I offer the following table to indicate what I might recommend for a student to read, according to his/her areas of interest, after this text is completed.

Projected area of interest	My suggestion as to what to read next, starting from a knowledge of this text.
Algebra	S. MacLane, *Homology*
Topology	A. Dold, *Lectures on Algebraic Topology*
Differential geometry	Steenrod, *The Topology of Fibre Bundles*
Analysis	"Topics" in J. Kelley's book, or in Bourbaki, particularly concerning complete metric spaces, and topological vector spaces.

A final word about problems. As I stated in the Preface, my feeling for this book was to supply a minimal list of problems, and to recommend that a student try and possibly work them *all*. All experts are unanimous in that one must work a certain number of such exercises to develop a legitimate understanding of any subject. Any student who wants to go on in mathematics ought to seek out some more advanced problems than those given here. I offer here my ideas on where to look for such problems.

a) *Point-set topology.* A general source of problems is Kelley. Bourbaki has a comprehensive list, working up to considerably difficult ones.

b) *Manifolds.* In general, Dold or Helgason. For surfaces, Massey.

c) *Covering spaces.* Massey.

d) *Fundamental group.* Seifert-Threlfall, Hilton-Wylie, or Massey.

Good Luck!

Bibliography

(References to foundations may be found at the end of Chapter 1.)

Atiyah, M. F.: *K-Theory*, Menlo Park, Calif.: W. A. Benjamin Inc., 1967.
Bartle, R., *The Elements of Real Analysis*, New York: Wiley-Interscience, 1967.
Bourbaki, N.: Many volumes published by Hermann et Cie, Paris. (English translations of some are available from Addison-Wesley Publishing Co., Inc., Reading, Mass.)
Dold, A.: *Lectures on Algebraic Topology*, New York: Springer-Verlag, 1972.
Dugundji, J.: *Topology*, Rockleigh, N. J.: Allyn & Bacon, 1966.
Eilenberg, S., and Steenrod, N.: *Foundations of Algebraic Topology*, Princeton, N. J.: Princeton University Press, 1952.
Gaal, S.: *Point-Set Topology*, New York: Academic Press, Inc., 1964.
Helgason, S.: *Differential Geometry and Symmetric Spaces*, New York: Academic Press, 1962.
Hilton, P. J., and Wylie, S.: *Homology Theory*, Cambridge, England: Cambridge University Press, 1960.
Hocking, J. G., and Young, G. S.: *Topology*, Reading, Mass.: Addison-Wesley Publishing Co., Inc., 1961.
Hu, S. T.: *Homotopy Theory*, New York: Academic Press, Inc., 1959.
Hurewicz, W., and Wallman, H.: *Dimension Theory*, Princeton, N. J.: Princeton University Press, 1948.
Kelley, J. L.: *General Topology*, New York: Van Nostrand Reinhold Co., 1955.
Kurosh, A. G.: *Group Theory*, Bronx, N. Y.: Chelsea Publishing Co., 1960.
Landau, E.: *Foundations of Analysis*, Bronx, N. Y.: Chelsea Publishing Co., 1951.
MacLane, S.: *Homology*, New York: Springer-Verlag, 1963.
Massey, W. S.: *Algebraic Topology: An Introduction*, New York: Harcourt Brace Jovanovich, Inc., 1967.
Maunder, C. R. F.: *Introduction to Algebraic Topology*, New York: Van Nostrand Reinhold Co., 1970.
Milnor, J. W.: *Morse Theory*, Princeton, N. J.: Princeton University Press, 1963.
Mitchell, B.: *Theory of Categories*, New York, Academic Press, Inc., 1965.
Montgomery, D., and Zippin, L.: *Topological Transformation Groups*, New York: Interscience, 1955.
Moore, R. L.: *Foundations of Point Set Theory*, Rev. Ed., Providence, R. I.: American Mathematics Society, 1962.
Pontryagin, L. S.: *Topological Groups*, Ed. 2, New York: Gordon & Breach, 1966.
Schubert, H.: *Katagorien*, 2 vols., New York: Springer-Verlag, 1970. (Translation: *Categories*, by J. Gray, 1972.)
Seifert, H., and Threlfall, W.: *Lehrbuch der Topologie*, Bronx, N. Y.: Chelsea Publishing Co., 1947.
Spanier, E. H.: *Algebraic Topology*, New York: McGraw-Hill Book Co., 1966.
Steenrod, N. E.: *The Topology of Fibre Bundles*, Princeton, N. J.: Princeton University Press, 1951.
Whitney, H.: *Geometric Integration Theory*, Princeton, N. J.: Princeton University Press, 1957.

Index of the most common symbols

$\Rightarrow$, 2
$\in$, 3
$\subseteq$, 4
$\cup$, 4
$\cap$, 4
ϕ, 5
f^{-1}, 5
$\underset{\alpha}{\times} X_\alpha$, 9
E^n (or $\mathbb{R}^n$), 16
$\langle X, Y \rangle$, 18
$\|X\|$, 19
H, 25
A^i, 37
A^e, 37
A^b, 37
$\bar{A}$, 38

A', 41
1_X, 14, 48
$f^{-1}(A)$, 31, 48
$f(A)$, 68
π_β, 50
$X/\sim$, 93
P^n, 112
$\Omega(X, x_0)$, 140
$\alpha \sim \beta$, 141
$\pi_1(X, x_0)$, 142
$\tilde{X}$, 157
(U, α), 174
σ^n, 183
b, 182
$St(v)$, 188
K^n, 187
$E_1(X, x_0)$, 193

Solutions to Selected Problems

Pages 20–21:

1. If one interval lies within the other, the result is clear. If $a < c < b < d$, then
$$\{x \mid a \leq x \leq b\} \cap \{x \mid c \leq x \leq d\}$$
equals $\{x \mid c \leq x \leq b\}$.
2. Take repeated mid-points.
3. $0 \leq \frac{1}{n} < \epsilon$ is equivalent to $n > \frac{1}{\epsilon} > 0$. Choose $a = 1$, $b = \frac{1}{\epsilon}$.
5. Draw a picture. Y is a unit-vector perpendicular to X.
7. The law of cosines for a triangle with sides of length A, B, C and α the angle between A and B says
$$C^2 = A^2 + B^2 - 2AB \cos \alpha.$$
Solve for α and interpret.
8. In C^1, put $Z = X$. Then interchange the roles of X and Y.

Page 23:

1. $x^2 + y^2 = 0$, etc.
2. A point on the line segment from X to Y has the form $tX + (1-t)Y$, where $0 \leq t \leq 1$. One calculates the distance from such a point to $\mathcal{C}$.
4. $\mathcal{C}$ is on a segment connecting two points on $S_R^{n-1}(\mathcal{C})$, but $\mathcal{C}$ does not lie on $S_R^{n-1}(\mathcal{C})$.
5. If you have located n points in E^{n-1}, whose distances from the origin are equal and whose mutual distances are 1, then it is an easy calculation to find an $(n+1)$-point on the last coordinate axis, which is at distance 1 from the previous n point.

Solutions to Selected Problems

Pages 33–34:

1. In the second case, note that the important thing to prove is
$$d_2(x, z) \leq d_2(x, y) + d_2(y, z).$$
If neither number on the right is 1 or smaller, then the right is just $d(x, y) + d(y, z) \geq d(x, z)$. But $d(x, z) \geq d_2(x, z)$. If one or more numbers on the right are 1, the sum is at least 1. Then it is automatically bigger than $d_2(x, y)$.

3. If $x > 0$, $y > 0$ and $xy < 1$, the distance from (x, y) to the positive x-axis, to the positive y-axis and to the curve $xy = 1$ are all positive numbers. The open ball around (x, y) of radius the minimum of these three positive numbers will be entirely in the set. This shows that the set is open.

4. A) $d_1((0, 1), (0, 2)) = 0$, showing that condition (A) of Definition 2.8 is violated.

5. B) If $x = 1$, $y = 0$, for example, then an open ball around (x, y), of positive radius, will contain points outside the region. Draw a picture.

6. Show that the inverse image of the open interval $\{x| -\epsilon < x < \epsilon\}$ is the open ball of radius $\epsilon > 0$ about the point x_0.

10. Choose an interval $\{x | a \leq x \leq b\}$ where $a \in \mathbb{R} - 0$, $b \in 0$. Let Z be the least upper bound of these points in this interval which belong to $\mathbb{R} - 0$. If on the other hand, $a \in 0$, $b \in \mathbb{R} - 0$, repeat the argument using a greatest lower bound.

Pages 42–43:

1. Apart from the discrete and the indiscrete topologies, there are the topologies with only one, non-trivial open set such as $\{a\}$ etc., and the topologies with three non-trivial open sets, such as $\{a\}$, $\{b\}$, $\{a, b\}$, and the topologies with only one non-trivial open set consisting of two elements, such as $\{a, b\}$. In addition, there are topologies with open sets of the form $\{a\}$, $\{a, b\}$. In the above, one may also—if desired—add an open set $\{c\}$.

2. $A^i = \langle 0, 1 \rangle \cup \langle 3, 4 \rangle$. $A^b = \{0\} \cup \{1\} \cup \{3\} \cup \{4\}$.

4. B) If x is an accumulation point of A, it is also an accumulation point of the larger B.

5. Let $A = [0, 1]$, $B = \mathbb{R}$. Then $B^b = \phi$.

8. Any open set around a real number x will contain both rational and irrational numbers.

9. $A^i = \phi$. $A^e = E^2 - A$. $A^b = A$, $A' = A$.

Page 47:

2. First calculate the intersection of a finite number of such sets.
3. If a single point is closed, then a finite union of points is also closed.

Pages 53–54:

1. b) $f^{-1}([-1, +1\rangle) = \langle-1, +1]$. This cannot be open, because if a basis set contained $+1$, then it would contain some points bigger than $+1$.
3. If B is closed in X, $X - B$ is open on X. $A - B = A \cap (X - B)$ is a relatively open set. Conversely, if B is closed in A, $A - B = A \cap 0$ for some open set in X. 0 may be assumed to contain $X - A$, then $X - B = (A - B) \cup (X - A) = (A \cap 0) \cup ((X - A) \cap 0) = 0$.
5. The inverse images of complementary sets are complementary.
7. A countable set with the discrete topology is an easy example.
8. If X is the union of two disjoint open sets, then Y is the union of their images under the homeomorphism.
11. $\tan(x)$ is a homeomorphism between $\langle -\frac{\pi}{2}, \frac{\pi}{2} \rangle$ and the entire real line.

Pages 67–68:

2. Each point is closed, so every subset must be closed.
3. Every open set contains all but a finite number of points.
5. If A and B are closed, $A \cap B$ is a closed subset of A.
6. Every subset of a finite space must be compact.
8. The sets $f^{-1}(\langle n, n+2 \rangle)$ are open sets which cover X.
9. One class of examples would be built from the same set, but X has the discrete topology.
11. Since f is now 1-1 and onto, the previous exercise will show that f sends open sets to open sets.
12. a) If a point x is at a distance of at least 1 from every point of the set S, x cannot be in the closure of S.
    b) By part a), the closure of a bounded set is compact. By problem 7, the image of this set is compact. Use Theorem 4.1.

Page 71:

1. A product of compact spaces would be compact but the real numbers are not.
3. Use the definition of an open set in a product and the fact that Z has a non-countable number of factors.

Solutions to Selected Problems

Page 76:

1. If $X = O_1 \cup O_2$, with these sets open and disjoint, set $f(x) = 0$ if $x \in O_1$, 1 otherwise.
2. The inverse image of disjoint open sets is made up of disjoint open sets.
3. Set $g(x) = x - f(x)$. $g: I \to [-1, 1]$. We may assume $g(0) < 0$ and $g(1) > 0$, or otherwise there is a fixed-point. Use problem 2 to show that there must be a c with $g(c) = 0$, thus $c = f(c)$.
5. One can do this with subsets of the circle $S^1 = \{(x, y) | x^2 + y^2 = 1\}$.
6. Show that any point can be connected to the north or to the south pole by an arc or path. Hence, any two points can be so connected.
8. Begin with closed intervals.

Pages 84–85:

2. Use Theorem 4.3 and Proposition 5.5. A metric space would satisfy the first axiom of countability. Let S be the subset of our space which is 0 in every coordinate except finitely many and 1 in those finitely many coordinates. The point which is 0 in each coordinate is an accumulation point of S, so S is not closed. But every sequence which is 1 on increasing finite subsets does not converge to a point in S.
3. A subspace of a metric space is a metric space.
4. Consider the possibility that X is discrete.
9. Two disjoint closed subsets of Y are disjoint closed subsets of X.

Page 91:

1. Different points are disjoint closed subsets.
3. Try a trivial topology on the infinite set.
4. Look at the metric in example 3 following Definition 2.8. The topology is discrete, but it is impossible that an infinite discrete space would satisfy the second axiom.

Page 97:

1. Recall that a continuous, 1-1, onto map between compact Hausdorff spaces is a homeomorphism.
2. An open neighborhood in $\mathbb{R}$ contains rationals and irrationals.
3. Use Definition 6.1.
6. A basis open set in a product is restricted to lie in prescribed open sets in a *finite* number of coordinates.

Solutions to Selected Problems 213

Page 104:

1. Use Tietze's theorem on each coordinate.
2. Let $A = \langle -\infty, 0 \rangle \cup \langle 0, +\infty \rangle \subseteq \mathbb{R}$, and select a suitable continuous function.
3. The image of a connected arc is a connected arc in I. Use Urysohn's lemma. Note also that a space with the indiscrete topology is arcwise connected.
4. The complements of single rational numbers are open. What is their intersection?
6. If it were, it would be a complete metric space of the first category.
8. The complement of a dense open set O_n is a nowhere dense closed set.

Page 115:

1. If O_1 and O_2 are homeomorphic to $\mathbb{R}^n$ and $\mathbb{R}^m$, then $O_1 \times O_2$ is homeomorphic to $\mathbb{R}^{n+m}$.
2. If a can be connected to b by a path, and b can be connected to c by a path, then a can be connected to c by a path. For example, if $\phi(0) = a$, $\phi(1) = b$, $\psi(0) = b$ and $\psi(1) = c$, then one may define a path by

$$\rho(t) = \begin{cases} \phi(2t), & \text{if } 0 \leq t \leq 1/2 \\ \psi(2t-1), & \text{if } \tfrac{1}{2} \leq t \leq 1. \end{cases}$$

Note also that any pair of points in $\mathbb{R}^n$ may be connected by a path.

3. $GL(n, \mathbb{R})$ is an open set in the space of all $n \times n$ matrices, which itself is homeomorphic to $\mathbb{R}^{n^2}$.
4. A multiplication may be defined by the formula $(x_1, y_1) \cdot (x_2, y_2) = (x_1 x_2, y_1 y_2)$.

Pages 136–137:

1. The methods of the proof of Theorem 7.2 apply. For example, $abcc^{-1}b^{-1}a^{-1}$ is equivalent to $abb^{-1}a$, by the reduction step 1).
2. Show explicitly that both the torus and the Klein bottle have subsets which are homeomorphic to S^1 whose complements are connected.
4. Show that a 2-manifold, which is differentiable, remains a differentiable manifold when a handle, or a cross-cap, is attached.

Page 154:

3. If f is homotopic to the constant map, there is $F: S^1 \times I \to Y$ so that $F(S^1 \times \{1\})$ is the constant map. F passes to a map of the quotient

$S^1 \times I/(a, 1) \sim (b, 1)$, which is easily seen to be homeomorphic to the space D^2.

4. A map which represents $\{\alpha\} + j_\#(\{\alpha\})$ will send the first half of the interval around the entire circle in one direction, and the second half of the interval back around in the other direction. The proof is very much the same as the proof that $\{\alpha\} \cdot \{\alpha\}^{-1} = e$.

Pages 178–179:

1. Verify that any set in the torus, which has diameter less than $\frac{1}{4}$ (have each parameter if the torus varies from 0 to 1), has the property that its inverse image consists of an infinite family of disjoint, homeomorphic copies. These copies are naturally in 1-1 correspondence with the group $Z \times Z$.

3. A covering space is arcwise connected by definition. If $p^{-1}(U) = \cup_\alpha V_\alpha$, then $p^{-1}(U \cap A) = \cup_\alpha V_\alpha \cap p^{-1}(A)$.

4. The fundamental group of a contractible space is trivial.

5. Choose e as the base point. The formula $(\alpha \cdot \beta)(t) = \alpha(t) \cdot \beta(t)$ defines a multiplication on paths beginning at e. Verify that the product in the topological group corresponds to multiplication in the fundamental group (recall $\pi_1(X \times Y, (x_o, y_o)) = \pi_1(X, x_o) \times \pi_1(Y, y_o)$). Then show the multiplication of paths, described above, is the equivalence relation used in Theorem 2.2.

Pages 201–202:

2. The paths which traverse each separate loop of the figure eight represent generators of the fundamental group. They do not commute with each other, so that the fundamental group is not abelian.

3. The generators may be taken to be the loops which go around in the two different directions. Theorem 10.2 shows that they do not commute.

5. This is a special case of problem 4.

6. Use Theorem 10.2. Since $A \cap B$ is presumed connected and $\pi_1(A \cap B) = 0$, all portions of the edge path in $A \cap B$ are basically equivalent.

7 and 8. S^2 or S^n breaks up into 2 contractible subcomplexes which meet on the equator.

9. There are five subgroups of $Z_2 \times Z_2$, including trivial ones.

Index

Accumulation point, 41
Archimedian order, 16
Arcwise connected, 75
Associative law, 113, 146
Axiom of choice, 10
Axioms of countability, 77

Baire category, 102
Ball, 22
Barycenter, 182
Barycentric subdivision, 186
Base point, 140
Basis, 44
Bolzano-Weierstrass property, 63
Borsuk-Ulam theorem, 178
Boundary (frontier), 37
Boundary (of a simplex), 184
Brouwer fixed-point theorem, 139

Cartesian product, 9
Category, 13
Category (Baire), 101
Cauchy sequence, 97
Cauchy-Schwartz inequality, 18
Closed interval, 22
Closed set, 37
Closure, 38
Coarser topology, 46
Commutative diagram, 95
Commutative law, 113
Compact, 57
Compactification (one point), 96
Complete, 101
Complex (simplicial), 124, 183
Cone, 110
Connected set, 72
Continuous, 27, 48
Contractible, 149
Converge (convergent sequence), 80
Convex, 22
Cover (open sets), 57
Covering homotopy, 161
Covering path, 159
Covering space, 157
Cross cap, 128

Deck transformation, 169
Dense, 78
Derived set, 41
Difference, 4
Differentiable manifold, 115
Dimension, 184
Directed set, 11
Disc, 116
Disconnected, 72
Discrete topology, 36
Distance function, 24
Distributative law, 15

Edge-path group, 193
Element, 3
Empty set, 5
Equivalence (edge-path), 193
Equivalence (relation), 93
Equivalent topologies, 47
Euclidean space, 16
Extension, 99
Exterior, 37

Face, 182
Fibre bundle, 205
Finer topology, 46
Finite intersection property, 63
First axiom of countability, 77
Fixed pont, 139
Free group, 198
Frontier (boundary), 37
Function, 5
Functor, 14
Fundamental group, 142

Graph, 184
Group
 finitely generated, 196
 free, 198
 fundamental, 142
 Lie, 114
 topological, 113

H space, 154
Handle, 128
Hausdorff space, 55

216 Index

Heine-Borel theorem, 63
Hilbert space, 25
Homeomorphism, 51
Homomorphism, 138
Homotopic maps, 141
Homotopy, 141
Homotopy class, 141

Identification (of edges), 117
Identity map, 13
Imbedding, 110
Inclusion, 49
Indiscrete, 36
Interior, 37
Intersection, 4
Invariance of dimension, 76
Inverse element, 114, 145
Inverse map, 5
Inverse path, 145
Isometric, 35
Isomorphic, 147

Klein bottle, 120

Least upper bound (supremum), 16
Lebesgue number, 190
Lie group, 114
Limit, 98
Locally arcwise connected, 156
Locally compact, 95
Loop, 140

Manifold, 105
Map, 5
Member, 3
Mesh, 186
Metric, 24
Metric space, 24, 88
Möbius band, 118
Multiplication (in a group), 113
Multiplication of loops, 141

Neighborhood, 43
Normal space, 81
Nowhere dense, 101
Null homotopic, 145

One-point compactification, 96
One-to-one, 5
Onto, 5
Open cover, 57
Open interval, 22
Open set, 28, 35
Open star, 188
Order, 11, 15

Paracompact, 204
Partially ordered, 10
Partition of unity, 108
Path, 75, 148
Path connected (arcwise connected), 75
Product, 9
Product (direct), 151
Product topology, 50
Projection, 50
Projective space, 112

Quotient, 92
Quotient space, 92
Quotient topology, 92

Rational numbers, 15
Real numbers, 15, 16
Relative topology, 49

Second axiom of countability, 77
Semi-locally simply connected, 171
Separable, 78
Sequence, 80
Set, 3
Simplex, 123, 181
Simplicial complex, 124, 183
Simply connected, 171
Sheets, 165
Skeleton, 192
Sphere, 22
Star, 138
Subbasis, 45
Subcomplex, 124
Subdivision, 186
Subset, 4
Subspace topology (relative topology), 49
Surface, 116

Tietze theorem, 99
Topological group, 113
Topological space, 35
Topology
  coarser, 46
  discrete, 36
  finer, 46
  indiscrete, 36
  metric, 28
  product, 50
  quotient, 92
  relative, 49
  weak (topology on a complex), 184
Torus, 120
Tychonoff theorem, 68

Uniformly convergent, 97

Union, 4
Unit interval, 99, 140
Universal covering space, 176
Urysohn's lemma, 85
Urysohn's metrization theorem, 88

Vector space, 17

Vertex, 123

Weak topology (of complex), 184
Well-ordered set, 11
Well-ordering principle, 11

Zorn's lemma, 11

A CATALOG OF SELECTED
DOVER BOOKS
IN SCIENCE AND MATHEMATICS

A CATALOG OF SELECTED
DOVER BOOKS
IN SCIENCE AND MATHEMATICS

QUALITATIVE THEORY OF DIFFERENTIAL EQUATIONS, V.V. Nemytskii and V.V. Stepanov. Classic graduate-level text by two prominent Soviet mathematicians covers classical differential equations as well as topological dynamics and ergodic theory. Bibliographies. 523pp. 5⅜ × 8½. 65954-2 Pa. $10.95

MATRICES AND LINEAR ALGEBRA, Hans Schneider and George Phillip Barker. Basic textbook covers theory of matrices and its applications to systems of linear equations and related topics such as determinants, eigenvalues and differential equations. Numerous exercises. 432pp. 5⅜ × 8½. 66014-1 Pa. $10.95

QUANTUM THEORY, David Bohm. This advanced undergraduate-level text presents the quantum theory in terms of qualitative and imaginative concepts, followed by specific applications worked out in mathematical detail. Preface. Index. 655pp. 5⅜ × 8½. 65969-0 Pa. $13.95

ATOMIC PHYSICS (8th edition), Max Born. Nobel laureate's lucid treatment of kinetic theory of gases, elementary particles, nuclear atom, wave-corpuscles, atomic structure and spectral lines, much more. Over 40 appendices, bibliography. 495pp. 5⅜ × 8½. 65984-4 Pa. $12.95

ELECTRONIC STRUCTURE AND THE PROPERTIES OF SOLIDS: The Physics of the Chemical Bond, Walter A. Harrison. Innovative text offers basic understanding of the electronic structure of covalent and ionic solids, simple metals, transition metals and their compounds. Problems. 1980 edition. 582pp. 6⅛ × 9¼. 66021-4 Pa. $15.95

BOUNDARY VALUE PROBLEMS OF HEAT CONDUCTION, M. Necati Özisik. Systematic, comprehensive treatment of modern mathematical methods of solving problems in heat conduction and diffusion. Numerous examples and problems. Selected references. Appendices. 505pp. 5⅜ × 8½. 65990-9 Pa. $12.95

A SHORT HISTORY OF CHEMISTRY (3rd edition), J.R. Partington. Classic exposition explores origins of chemistry, alchemy, early medical chemistry, nature of atmosphere, theory of valency, laws and structure of atomic theory, much more. 428pp. 5⅜ × 8½. (Available in U.S. only) 65977-1 Pa. $10.95

A HISTORY OF ASTRONOMY, A. Pannekoek. Well-balanced, carefully reasoned study covers such topics as Ptolemaic theory, work of Copernicus, Kepler, Newton, Eddington's work on stars, much more. Illustrated. References. 521pp. 5⅜ × 8½. 65994-1 Pa. $12.95

PRINCIPLES OF METEOROLOGICAL ANALYSIS, Walter J. Saucier. Highly respected, abundantly illustrated classic reviews atmospheric variables, hydrostatics, static stability, various analyses (scalar, cross-section, isobaric, isentropic, more). For intermediate meteorology students. 454pp. 6⅛ × 9¼. 65979-8 Pa. $14.95

CATALOG OF DOVER BOOKS

NUMERICAL METHODS FOR SCIENTISTS AND ENGINEERS, Richard Hamming. Classic text stresses frequency approach in coverage of algorithms, polynomial approximation, Fourier approximation, exponential approximation, other topics. Revised and enlarged 2nd edition. 721pp. 5⅜ × 8½.
65241-6 Pa. $14.95

THEORETICAL SOLID STATE PHYSICS, Vol. I: Perfect Lattices in Equilibrium; Vol. II: Non-Equilibrium and Disorder, William Jones and Norman H. March. Monumental reference work covers fundamental theory of equilibrium properties of perfect crystalline solids, non-equilibrium properties, defects and disordered systems. Appendices. Problems. Preface. Diagrams. Index. Bibliography. Total of 1,301pp. 5⅜ × 8½. Two volumes.   Vol. I 65015-4 Pa. $14.95
Vol. II 65016-2 Pa. $14.95

OPTIMIZATION THEORY WITH APPLICATIONS, Donald A. Pierre. Broadspectrum approach to important topic. Classical theory of minima and maxima, calculus of variations, simplex technique and linear programming, more. Many problems, examples. 640pp. 5⅜ × 8½.
65205-X Pa. $14.95

THE CONTINUUM: A Critical Examination of the Foundation of Analysis, Hermann Weyl. Classic of 20th-century foundational research deals with the conceptual problem posed by the continuum. 156pp. 5⅜ × 8½.   67982-9 Pa. $5.95

ESSAYS ON THE THEORY OF NUMBERS, Richard Dedekind. Two classic essays by great German mathematician: on the theory of irrational numbers; and on transfinite numbers and properties of natural numbers. 115pp. 5⅜ × 8½.
21010-3 Pa. $4.95

THE FUNCTIONS OF MATHEMATICAL PHYSICS, Harry Hochstadt. Comprehensive treatment of orthogonal polynomials, hypergeometric functions, Hill's equation, much more. Bibliography. Index. 322pp. 5⅜ × 8½.   65214-9 Pa. $9.95

NUMBER THEORY AND ITS HISTORY, Oystein Ore. Unusually clear, accessible introduction covers counting, properties of numbers, prime numbers, much more. Bibliography. 380pp. 5⅜ × 8½.   65620-9 Pa. $9.95

THE VARIATIONAL PRINCIPLES OF MECHANICS, Cornelius Lanczos. Graduate level coverage of calculus of variations, equations of motion, relativistic mechanics, more. First inexpensive paperbound edition of classic treatise. Index. Bibliography. 418pp. 5⅜ × 8½.   65067-7 Pa. $11.95

MATHEMATICAL TABLES AND FORMULAS, Robert D. Carmichael and Edwin R. Smith. Logarithms, sines, tangents, trig functions, powers, roots, reciprocals, exponential and hyperbolic functions, formulas and theorems. 269pp. 5⅜ × 8½.   60111-0 Pa. $6.95

THEORETICAL PHYSICS, Georg Joos, with Ira M. Freeman. Classic overview covers essential math, mechanics, electromagnetic theory, thermodynamics, quantum mechanics, nuclear physics, other topics. First paperback edition. xxiii + 885pp. 5⅜ × 8½.   65227-0 Pa. $19.95

HANDBOOK OF MATHEMATICAL FUNCTIONS WITH FORMULAS, GRAPHS, AND MATHEMATICAL TABLES, edited by Milton Abramowitz and Irene A. Stegun. Vast compendium: 29 sets of tables, some to as high as 20 places. 1,046pp. 8 × 10½. 61272-4 Pa. $24.95

MATHEMATICAL METHODS IN PHYSICS AND ENGINEERING, John W. Dettman. Algebraically based approach to vectors, mapping, diffraction, other topics in applied math. Also generalized functions, analytic function theory, more. Exercises. 448pp. 5⅜ × 8¼. 65649-7 Pa. $9.95

A SURVEY OF NUMERICAL MATHEMATICS, David M. Young and Robert Todd Gregory. Broad self-contained coverage of computer-oriented numerical algorithms for solving various types of mathematical problems in linear algebra, ordinary and partial, differential equations, much more. Exercises. Total of 1,248pp. 5⅜ × 8½. Two volumes. Vol. I 65691-8 Pa. $14.95
Vol. II 65692-6 Pa. $14.95

TENSOR ANALYSIS FOR PHYSICISTS, J.A. Schouten. Concise exposition of the mathematical basis of tensor analysis, integrated with well-chosen physical examples of the theory. Exercises. Index. Bibliography. 289pp. 5⅜ × 8½. 65582-2 Pa. $8.95

INTRODUCTION TO NUMERICAL ANALYSIS (2nd Edition), F.B. Hildebrand. Classic, fundamental treatment covers computation, approximation, interpolation, numerical differentiation and integration, other topics. 150 new problems. 669pp. 5⅜ × 8½. 65363-3 Pa. $15.95

INVESTIGATIONS ON THE THEORY OF THE BROWNIAN MOVEMENT, Albert Einstein. Five papers (1905–8) investigating dynamics of Brownian motion and evolving elementary theory. Notes by R. Fürth. 122pp. 5⅜ × 8½. 60304-0 Pa. $4.95

CATASTROPHE THEORY FOR SCIENTISTS AND ENGINEERS, Robert Gilmore. Advanced-level treatment describes mathematics of theory grounded in the work of Poincaré, R. Thom, other mathematicians. Also important applications to problems in mathematics, physics, chemistry and engineering. 1981 edition. References. 28 tables. 397 black-and-white illustrations. xvii + 666pp. 6⅛ × 9¼. 67539-4 Pa. $16.95

AN INTRODUCTION TO STATISTICAL THERMODYNAMICS, Terrell L. Hill. Excellent basic text offers wide-ranging coverage of quantum statistical mechanics, systems of interacting molecules, quantum statistics, more. 523pp. 5⅜ × 8½. 65242-4 Pa. $12.95

ELEMENTARY DIFFERENTIAL EQUATIONS, William Ted Martin and Eric Reissner. Exceptionally clear, comprehensive introduction at undergraduate level. Nature and origin of differential equations, differential equations of first, second and higher orders. Picard's Theorem, much more. Problems with solutions. 331pp. 5⅜ × 8½. 65024-3 Pa. $8.95

STATISTICAL PHYSICS, Gregory H. Wannier. Classic text combines thermodynamics, statistical mechanics and kinetic theory in one unified presentation of thermal physics. Problems with solutions. Bibliography. 532pp. 5⅜ × 8½. 65401-X Pa. $12.95

CATALOG OF DOVER BOOKS

ORDINARY DIFFERENTIAL EQUATIONS, Morris Tenenbaum and Harry Pollard. Exhaustive survey of ordinary differential equations for undergraduates in mathematics, engineering, science. Thorough analysis of theorems. Diagrams. Bibliography. Index. 818pp. 5⅜ × 8½. 64940-7 Pa. $16.95

STATISTICAL MECHANICS: Principles and Applications, Terrell L. Hill. Standard text covers fundamentals of statistical mechanics, applications to fluctuation theory, imperfect gases, distribution functions, more. 448pp. 5⅜ × 8½. 65390-0 Pa. $11.95

ORDINARY DIFFERENTIAL EQUATIONS AND STABILITY THEORY: An Introduction, David A. Sánchez. Brief, modern treatment. Linear equation, stability theory for autonomous and nonautonomous systems, etc. 164pp. 5⅜ × 8¼. 63828-6 Pa. $5.95

THIRTY YEARS THAT SHOOK PHYSICS: The Story of Quantum Theory, George Gamow. Lucid, accessible introduction to influential theory of energy and matter. Careful explanations of Dirac's anti-particles, Bohr's model of the atom, much more. 12 plates. Numerous drawings. 240pp. 5⅜ × 8½. 24895-X Pa. $6.95

THEORY OF MATRICES, Sam Perlis. Outstanding text covering rank, non-singularity and inverses in connection with the development of canonical matrices under the relation of equivalence, and without the intervention of determinants. Includes exercises. 237pp. 5⅜ × 8½. 66810-X Pa. $7.95

GREAT EXPERIMENTS IN PHYSICS: Firsthand Accounts from Galileo to Einstein, edited by Morris H. Shamos. 25 crucial discoveries: Newton's laws of motion, Chadwick's study of the neutron, Hertz on electromagnetic waves, more. Original accounts clearly annotated. 370pp. 5⅜ × 8½. 25346-5 Pa. $10.95

INTRODUCTION TO PARTIAL DIFFERENTIAL EQUATIONS WITH APPLICATIONS, E.C. Zachmanoglou and Dale W. Thoe. Essentials of partial differential equations applied to common problems in engineering and the physical sciences. Problems and answers. 416pp. 5⅜ × 8½. 65251-3 Pa. $10.95

BURNHAM'S CELESTIAL HANDBOOK, Robert Burnham, Jr. Thorough guide to the stars beyond our solar system. Exhaustive treatment. Alphabetical by constellation: Andromeda to Cetus in Vol. 1; Chamaeleon to Orion in Vol. 2; and Pavo to Vulpecula in Vol. 3. Hundreds of illustrations. Index in Vol. 3. 2,000pp. 6⅛ × 9¼. 23567-X, 23568-8, 23673-0 Pa., Three-vol. set $41.85

CHEMICAL MAGIC, Leonard A. Ford. Second Edition, Revised by E. Winston Grundmeier. Over 100 unusual stunts demonstrating cold fire, dust explosions, much more. Text explains scientific principles and stresses safety precautions. 128pp. 5⅜ × 8½. 67628-5 Pa. $5.95

AMATEUR ASTRONOMER'S HANDBOOK, J.B. Sidgwick. Timeless, comprehensive coverage of telescopes, mirrors, lenses, mountings, telescope drives, micrometers, spectroscopes, more. 189 illustrations. 576pp. 5⅜ × 8¼. (Available in U.S. only) 24034-7 Pa. $9.95

CATALOG OF DOVER BOOKS

SPECIAL FUNCTIONS, N.N. Lebedev. Translated by Richard Silverman. Famous Russian work treating more important special functions, with applications to specific problems of physics and engineering. 38 figures. 308pp. 5⅜ × 8½.
60624-4 Pa. $8.95

OBSERVATIONAL ASTRONOMY FOR AMATEURS, J.B. Sidgwick. Mine of useful data for observation of sun, moon, planets, asteroids, aurorae, meteors, comets, variables, binaries, etc. 39 illustrations. 384pp. 5⅜ × 8¼. (Available in U.S. only)
24033-9 Pa. $8.95

INTEGRAL EQUATIONS, F.G. Tricomi. Authoritative, well-written treatment of extremely useful mathematical tool with wide applications. Volterra Equations, Fredholm Equations, much more. Advanced undergraduate to graduate level. Exercises. Bibliography. 238pp. 5⅜ × 8½.
64828-1 Pa. $7.95

POPULAR LECTURES ON MATHEMATICAL LOGIC, Hao Wang. Noted logician's lucid treatment of historical developments, set theory, model theory, recursion theory and constructivism, proof theory, more. 3 appendixes. Bibliography. 1981 edition. ix + 283pp. 5⅜ × 8½.
67632-3 Pa. $8.95

MODERN NONLINEAR EQUATIONS, Thomas L. Saaty. Emphasizes practical solution of problems; covers seven types of equations. ". . . a welcome contribution to the existing literature. . . ."—*Math Reviews.* 490pp. 5⅜ × 8½. 64232-1 Pa. $11.95

FUNDAMENTALS OF ASTRODYNAMICS, Roger Bate et al. Modern approach developed by U.S. Air Force Academy. Designed as a first course. Problems, exercises. Numerous illustrations. 455pp. 5⅜ × 8½.
60061-0 Pa. $9.95

INTRODUCTION TO LINEAR ALGEBRA AND DIFFERENTIAL EQUATIONS, John W. Dettman. Excellent text covers complex numbers, determinants, orthonormal bases, Laplace transforms, much more. Exercises with solutions. Undergraduate level. 416pp. 5⅜ × 8½.
65191-6 Pa. $10.95

INCOMPRESSIBLE AERODYNAMICS, edited by Bryan Thwaites. Covers theoretical and experimental treatment of the uniform flow of air and viscous fluids past two-dimensional aerofoils and three-dimensional wings; many other topics. 654pp. 5⅜ × 8½.
65465-6 Pa. $16.95

INTRODUCTION TO DIFFERENCE EQUATIONS, Samuel Goldberg. Exceptionally clear exposition of important discipline with applications to sociology, psychology, economics. Many illustrative examples; over 250 problems. 260pp. 5⅜ × 8½.
65084-7 Pa. $7.95

LAMINAR BOUNDARY LAYERS, edited by L. Rosenhead. Engineering classic covers steady boundary layers in two- and three-dimensional flow, unsteady boundary layers, stability, observational techniques, much more. 708pp. 5⅜ × 8½.
65646-2 Pa. $18.95

LECTURES ON CLASSICAL DIFFERENTIAL GEOMETRY, Second Edition, Dirk J. Struik. Excellent brief introduction covers curves, theory of surfaces, fundamental equations, geometry on a surface, conformal mapping, other topics. Problems. 240pp. 5⅜ × 8½.
65609-8 Pa. $8.95

CATALOG OF DOVER BOOKS

ROTARY-WING AERODYNAMICS, W.Z. Stepniewski. Clear, concise text covers aerodynamic phenomena of the rotor and offers guidelines for helicopter performance evaluation. Originally prepared for NASA. 537 figures. 640pp. 6⅛ × 9¼.
64647-5 Pa. $15.95

DIFFERENTIAL GEOMETRY, Heinrich W. Guggenheimer. Local differential geometry as an application of advanced calculus and linear algebra. Curvature, transformation groups, surfaces, more. Exercises. 62 figures. 378pp. 5⅜ × 8½.
63433-7 Pa. $8.95

INTRODUCTION TO SPACE DYNAMICS, William Tyrrell Thomson. Comprehensive, classic introduction to space-flight engineering for advanced undergraduate and graduate students. Includes vector algebra, kinematics, transformation of coordinates. Bibliography. Index. 352pp. 5⅜ × 8½. 65113-4 Pa. $8.95

A SURVEY OF MINIMAL SURFACES, Robert Osserman. Up-to-date, in-depth discussion of the field for advanced students. Corrected and enlarged edition covers new developments. Includes numerous problems. 192pp. 5⅜ × 8½.
64998-9 Pa. $8.95

ANALYTICAL MECHANICS OF GEARS, Earle Buckingham. Indispensable reference for modern gear manufacture covers conjugate gear-tooth action, gear-tooth profiles of various gears, many other topics. 263 figures. 102 tables. 546pp. 5⅜ × 8½.
65712-4 Pa. $14.95

SET THEORY AND LOGIC, Robert R. Stoll. Lucid introduction to unified theory of mathematical concepts. Set theory and logic seen as tools for conceptual understanding of real number system. 496pp. 5⅜ × 8¼. 63829-4 Pa. $12.95

A HISTORY OF MECHANICS, René Dugas. Monumental study of mechanical principles from antiquity to quantum mechanics. Contributions of ancient Greeks, Galileo, Leonardo, Kepler, Lagrange, many others. 671pp. 5⅜ × 8½.
65632-2 Pa. $14.95

FAMOUS PROBLEMS OF GEOMETRY AND HOW TO SOLVE THEM, Benjamin Bold. Squaring the circle, trisecting the angle, duplicating the cube: learn their history, why they are impossible to solve, then solve them yourself. 128pp. 5⅜ × 8½.
24297-8 Pa. $4.95

MECHANICAL VIBRATIONS, J.P. Den Hartog. Classic textbook offers lucid explanations and illustrative models, applying theories of vibrations to a variety of practical industrial engineering problems. Numerous figures. 233 problems, solutions. Appendix. Index. Preface. 436pp. 5⅜ × 8½. 64785-4 Pa. $10.95

CURVATURE AND HOMOLOGY, Samuel I. Goldberg. Thorough treatment of specialized branch of differential geometry. Covers Riemannian manifolds, topology of differentiable manifolds, compact Lie groups, other topics. Exercises. 315pp. 5⅜ × 8½.
64314-X Pa. $9.95

HISTORY OF STRENGTH OF MATERIALS, Stephen P. Timoshenko. Excellent historical survey of the strength of materials with many references to the theories of elasticity and structure. 245 figures. 452pp. 5⅜ × 8½. 61187-6 Pa. $11.95

CATALOG OF DOVER BOOKS

GEOMETRY OF COMPLEX NUMBERS, Hans Schwerdtfeger. Illuminating, widely praised book on analytic geometry of circles, the Moebius transformation, and two-dimensional non-Euclidean geometries. 200pp. 5⅜ × 8¼.
63830-8 Pa. $8.95

MECHANICS, J.P. Den Hartog. A classic introductory text or refresher. Hundreds of applications and design problems illuminate fundamentals of trusses, loaded beams and cables, etc. 334 answered problems. 462pp. 5⅜ × 8½. 60754-2 Pa. $9.95

TOPOLOGY, John G. Hocking and Gail S. Young. Superb one-year course in classical topology. Topological spaces and functions, point-set topology, much more. Examples and problems. Bibliography. Index. 384pp. 5⅜ × 8¼.
65676-4 Pa. $9.95

STRENGTH OF MATERIALS, J.P. Den Hartog. Full, clear treatment of basic material (tension, torsion, bending, etc.) plus advanced material on engineering methods, applications. 350 answered problems. 323pp. 5⅜ × 8½. 60755-0 Pa. $8.95

ELEMENTARY CONCEPTS OF TOPOLOGY, Paul Alexandroff. Elegant, intuitive approach to topology from set-theoretic topology to Betti groups; how concepts of topology are useful in math and physics. 25 figures. 57pp. 5⅜ × 8½.
60747-X Pa. $3.50

ADVANCED STRENGTH OF MATERIALS, J.P. Den Hartog. Superbly written advanced text covers torsion, rotating disks, membrane stresses in shells, much more. Many problems and answers. 388pp. 5⅜ × 8½. 65407-9 Pa. $9.95

COMPUTABILITY AND UNSOLVABILITY, Martin Davis. Classic graduate-level introduction to theory of computability, usually referred to as theory of recurrent functions. New preface and appendix. 288pp. 5⅜ × 8½. 61471-9 Pa. $7.95

GENERAL CHEMISTRY, Linus Pauling. Revised 3rd edition of classic first-year text by Nobel laureate. Atomic and molecular structure, quantum mechanics, statistical mechanics, thermodynamics correlated with descriptive chemistry. Problems. 992pp. 5⅜ × 8½. 65622-5 Pa. $19.95

AN INTRODUCTION TO MATRICES, SETS AND GROUPS FOR SCIENCE STUDENTS, G. Stephenson. Concise, readable text introduces sets, groups, and most importantly, matrices to undergraduate students of physics, chemistry, and engineering. Problems. 164pp. 5⅜ × 8½. 65077-4 Pa. $6.95

THE HISTORICAL BACKGROUND OF CHEMISTRY, Henry M. Leicester. Evolution of ideas, not individual biography. Concentrates on formulation of a coherent set of chemical laws. 260pp. 5⅜ × 8½. 61053-5 Pa. $6.95

THE PHILOSOPHY OF MATHEMATICS: An Introductory Essay, Stephan Körner. Surveys the views of Plato, Aristotle, Leibniz & Kant concerning propositions and theories of applied and pure mathematics. Introduction. Two appendices. Index. 198pp. 5⅜ × 8½. 25048-2 Pa. $7.95

THE DEVELOPMENT OF MODERN CHEMISTRY, Aaron J. Ihde. Authoritative history of chemistry from ancient Greek theory to 20th-century innovation. Covers major chemists and their discoveries. 209 illustrations. 14 tables. Bibliographies. Indices. Appendices. 851pp. 5⅜ × 8½. 64235-6 Pa. $18.95

CATALOG OF DOVER BOOKS

DE RE METALLICA, Georgius Agricola. The famous Hoover translation of greatest treatise on technological chemistry, engineering, geology, mining of early modern times (1556). All 289 original woodcuts. 638pp. 6¾ × 11.
60006-8 Pa. $18.95

SOME THEORY OF SAMPLING, William Edwards Deming. Analysis of the problems, theory and design of sampling techniques for social scientists, industrial managers and others who find statistics increasingly important in their work. 61 tables. 90 figures. xvii + 602pp. 5⅜ × 8½.
64684-X Pa. $15.95

THE VARIOUS AND INGENIOUS MACHINES OF AGOSTINO RAMELLI: A Classic Sixteenth-Century Illustrated Treatise on Technology, Agostino Ramelli. One of the most widely known and copied works on machinery in the 16th century. 194 detailed plates of water pumps, grain mills, cranes, more. 608pp. 9 × 12.
28180-9 Pa. $24.95

LINEAR PROGRAMMING AND ECONOMIC ANALYSIS, Robert Dorfman, Paul A. Samuelson and Robert M. Solow. First comprehensive treatment of linear programming in standard economic analysis. Game theory, modern welfare economics, Leontief input-output, more. 525pp. 5⅜ × 8½.
65491-5 Pa. $14.95

ELEMENTARY DECISION THEORY, Herman Chernoff and Lincoln E. Moses. Clear introduction to statistics and statistical theory covers data processing, probability and random variables, testing hypotheses, much more. Exercises. 364pp. 5⅜ × 8½.
65218-1 Pa. $9.95

THE COMPLEAT STRATEGYST: Being a Primer on the Theory of Games of Strategy, J.D. Williams. Highly entertaining classic describes, with many illustrated examples, how to select best strategies in conflict situations. Prefaces. Appendices. 268pp. 5⅜ × 8½.
25101-2 Pa. $7.95

MATHEMATICAL METHODS OF OPERATIONS RESEARCH, Thomas L. Saaty. Classic graduate-level text covers historical background, classical methods of forming models, optimization, game theory, probability, queueing theory, much more. Exercises. Bibliography. 448pp. 5⅜ × 8¼.
65703-5 Pa. $12.95

CONSTRUCTIONS AND COMBINATORIAL PROBLEMS IN DESIGN OF EXPERIMENTS, Damaraju Raghavarao. In-depth reference work examines orthogonal Latin squares, incomplete block designs, tactical configuration, partial geometry, much more. Abundant explanations, examples. 416pp. 5⅜ × 8¼.
65685-3 Pa. $10.95

THE ABSOLUTE DIFFERENTIAL CALCULUS (CALCULUS OF TENSORS), Tullio Levi-Civita. Great 20th-century mathematician's classic work on material necessary for mathematical grasp of theory of relativity. 452pp. 5⅜ × 8½.
63401-9 Pa. $9.95

VECTOR AND TENSOR ANALYSIS WITH APPLICATIONS, A.I. Borisenko and I.E. Tarapov. Concise introduction. Worked-out problems, solutions, exercises. 257pp. 5⅜ × 8¼.
63833-2 Pa. $7.95

CATALOG OF DOVER BOOKS

THE FOUR-COLOR PROBLEM: Assaults and Conquest, Thomas L. Saaty and Paul G. Kainen. Engrossing, comprehensive account of the century-old combinatorial topological problem, its history and solution. Bibliographies. Index. 110 figures. 228pp. 5⅜ × 8½. 65092-8 Pa. $6.95

CATALYSIS IN CHEMISTRY AND ENZYMOLOGY, William P. Jencks. Exceptionally clear coverage of mechanisms for catalysis, forces in aqueous solution, carbonyl- and acyl-group reactions, practical kinetics, more. 864pp. 5⅜ × 8½. 65460-5 Pa. $19.95

PROBABILITY: An Introduction, Samuel Goldberg. Excellent basic text covers set theory, probability theory for finite sample spaces, binomial theorem, much more. 360 problems. Bibliographies. 322pp. 5⅜ × 8½. 65252-1 Pa. $8.95

LIGHTNING, Martin A. Uman. Revised, updated edition of classic work on the physics of lightning. Phenomena, terminology, measurement, photography, spectroscopy, thunder, more. Reviews recent research. Bibliography. Indices. 320pp. 5⅜ × 8¼. 64575-4 Pa. $8.95

PROBABILITY THEORY: A Concise Course, Y.A. Rozanov. Highly readable, self-contained introduction covers combination of events, dependent events, Bernoulli trials, etc. Translation by Richard Silverman. 148pp. 5⅜ × 8¼.
63544-9 Pa. $5.95

AN INTRODUCTION TO HAMILTONIAN OPTICS, H. A. Buchdahl. Detailed account of the Hamiltonian treatment of aberration theory in geometrical optics. Many classes of optical systems defined in terms of the symmetries they possess. Problems with detailed solutions. 1970 edition. xv + 360pp. 5⅜ × 8½.
67597-1 Pa. $10.95

STATISTICS MANUAL, Edwin L. Crow, et al. Comprehensive, practical collection of classical and modern methods prepared by U.S. Naval Ordnance Test Station. Stress on use. Basics of statistics assumed. 288pp. 5⅜ × 8½.
60599-X Pa. $6.95

DICTIONARY/OUTLINE OF BASIC STATISTICS, John E. Freund and Frank J. Williams. A clear concise dictionary of over 1,000 statistical terms and an outline of statistical formulas covering probability, nonparametric tests, much more. 208pp. 5⅜ × 8½. 66796-0 Pa. $6.95

STATISTICAL METHOD FROM THE VIEWPOINT OF QUALITY CONTROL, Walter A. Shewhart. Important text explains regulation of variables, uses of statistical control to achieve quality control in industry, agriculture, other areas. 192pp. 5⅜ × 8½. 65232-7 Pa. $7.95

THE INTERPRETATION OF GEOLOGICAL PHASE DIAGRAMS, Ernest G. Ehlers. Clear, concise text emphasizes diagrams of systems under fluid or containing pressure; also coverage of complex binary systems, hydrothermal melting, more. 288pp. 6½ × 9¼. 65389-7 Pa. $10.95

STATISTICAL ADJUSTMENT OF DATA, W. Edwards Deming. Introduction to basic concepts of statistics, curve fitting, least squares solution, conditions without parameter, conditions containing parameters. 26 exercises worked out. 271pp. 5⅜ × 8½. 64685-8 Pa. $8.95

CATALOG OF DOVER BOOKS

CHALLENGING MATHEMATICAL PROBLEMS WITH ELEMENTARY SOLUTIONS, A.M. Yaglom and I.M. Yaglom. Over 170 challenging problems on probability theory, combinatorial analysis, points and lines, topology, convex polygons, many other topics. Solutions. Total of 445pp. 5⅜ × 8½. Two-vol. set.
Vol. I 65536-9 Pa. $7.95
Vol. II 65537-7 Pa. $6.95

FIFTY CHALLENGING PROBLEMS IN PROBABILITY WITH SOLUTIONS, Frederick Mosteller. Remarkable puzzlers, graded in difficulty, illustrate elementary and advanced aspects of probability. Detailed solutions. 88pp. 5⅜ × 8½.
65355-2 Pa. $4.95

EXPERIMENTS IN TOPOLOGY, Stephen Barr. Classic, lively explanation of one of the byways of mathematics. Klein bottles, Moebius strips, projective planes, map coloring, problem of the Koenigsberg bridges, much more, described with clarity and wit. 43 figures. 210pp. 5⅜ × 8½. 25933-1 Pa. $5.95

RELATIVITY IN ILLUSTRATIONS, Jacob T. Schwartz. Clear nontechnical treatment makes relativity more accessible than ever before. Over 60 drawings illustrate concepts more clearly than text alone. Only high school geometry needed. Bibliography. 128pp. 6⅛ × 9¼. 25965-X Pa. $6.95

AN INTRODUCTION TO ORDINARY DIFFERENTIAL EQUATIONS, Earl A. Coddington. A thorough and systematic first course in elementary differential equations for undergraduates in mathematics and science, with many exercises and problems (with answers). Index. 304pp. 5⅜ × 8½. 65942-9 Pa. $8.95

FOURIER SERIES AND ORTHOGONAL FUNCTIONS, Harry F. Davis. An incisive text combining theory and practical example to introduce Fourier series, orthogonal functions and applications of the Fourier method to boundary-value problems. 570 exercises. Answers and notes. 416pp. 5⅜ × 8½. 65973-9 Pa. $9.95

THE THEORY OF BRANCHING PROCESSES, Theodore E. Harris. First systematic, comprehensive treatment of branching (i.e. multiplicative) processes and their applications. Galton-Watson model, Markov branching processes, electron-photon cascade, many other topics. Rigorous proofs. Bibliography. 240pp. 5⅜ × 8½. 65952-6 Pa. $6.95

AN INTRODUCTION TO ALGEBRAIC STRUCTURES, Joseph Landin. Superb self-contained text covers "abstract algebra": sets and numbers, theory of groups, theory of rings, much more. Numerous well-chosen examples, exercises. 247pp. 5⅜ × 8½. 65940-2 Pa. $7.95

Prices subject to change without notice.
Available at your book dealer or write for free Mathematics and Science Catalog to Dept. GI, Dover Publications, Inc., 31 East 2nd St., Mineola, N.Y. 11501. Dover publishes more than 175 books each year on science, elementary and advanced mathematics, biology, music, art, literature, history, social sciences and other areas.